T0094197

'The blending of Business and OHS knowledge makes this book a must read for any aspiring or practising OHS professional who wants to make a real difference. I wish it had been available when I began my OHS career some 30 years ago'.
—*Dr. Dominic Cooper*, CFIOSH CPsychol, author of 'Improving Safety Culture: A Practical Guide', and 'Strategic Safety Culture Roadmap'

'An excellent resource with ample practical tips for OSH professionals to transform their engagement with senior management to achieve enhanced business performance. Ambitious OSH professionals will find themselves dipping into this superb book time and again for many years'.
—*Olumide Adeolu*, BEng (Hons) MSc MA CMIOSH CSP, Past Chair IOSH Qatar Branch

The 10 Step MBA for Safety and Health Practitioners

As an Occupational Safety and Health (OSH) practitioner have you ever wondered "How can I shape my career trajectory to reach a C-suite position in business?" Or perhaps – for those who do not aspire to positions of this nature: "How can I develop my ability to persuade and influence top management more effectively?" *The 10 Step MBA for Safety and Health Practitioners* answers these questions to enable you to achieve your personal and professional OSH goals.

Presented over 10 steps encompassing a typical MBA programme, a transformational model establishes key themes which are deemed critical in understanding the world of business to exert greater influence:

- **Strategic** – aligning OSH to the overall direction of a business and creating a lasting OSH purpose that all stakeholders can relate to
- **Cross-functional** – understanding the different parts of an organisation and integrating OSH within business functions and ways of working
- **Distinctive** – looking for creative new ways of presenting OSH data and information to generate interest and enthusiasm.

From strategy and leadership to organisational behaviour and human resource management, from marketing and brand management to interpersonal skills, this book shows you how to combine the best of your specialist knowledge with important business tools so you can embed OSH at the heart of your company. The book is an indispensable reference for OSH practitioners who want to make a positive change in their careers and become more effective in influencing and leading change.

Waddah S Ghanem Al Hashmi is Executive Director of EHSSQ and Corporate Affairs at the ENOC Group in Dubai, United Arab Emirates. Waddah is responsible for overseeing the development and implementation of best practices and standards for Legal affairs, Group Communications, EHS, Business Excellence & Quality, Sustainability, Security and Risk Management. Waddah holds a DBA in Corporate Governance and EHS Leadership from the University of Bradford's School of Management.

Rob Cooling is Director of Health, Safety, Quality and Environment at Expo 2020 Dubai. Rob is a Trustee of the Institution of Occupational Safety and Health (IOSH) and a Chartered Fellow of IOSH. Rob holds a PhD in health & safety law and management from Salford University, UK and an Executive MBA from London Business School, UK.

The 10 Step MBA for Safety and Health Practitioners

Waddah S Ghanem Al Hashmi
and Rob Cooling

Routledge
Taylor & Francis Group

LONDON AND NEW YORK

First published 2018
by Routledge
2 Park Square, Milton Park, Abingdon, Oxon OX14 4RN

and by Routledge
711 Third Avenue, New York, NY 10017

*Routledge is an imprint of the Taylor & Francis Group, an informa
business*

British Library Cataloguing-in-Publication Data
A catalogue record for this book is available from the British Library

Library of Congress Cataloguing-in-Publication Data
Names: Shihab Ghanem Al Hashmi, Waddah, author. | Cooling, Rob,
    author.
Title: The 10 step MBA for safety and health practitioners / Waddah S
    Ghanem Al Hashmi and Rob Cooling.
Other titles: Ten step MBA for safety and health practitioners
Description: Abingdon, Oxon; New York, NY: Routledge, 2018. |
    Includes bibliographical references and index.
Identifiers: LCCN 2017050773 | ISBN 9781138068667 (hardback) | ISBN
    9781138821965 (pbk.) | ISBN 9781315743011 (ebook)
Subjects: MESH: Occupational Health | Institutional Practice | Practice
    Management | Organizational Innovation | Social Responsibility
Classification: LCC RC967 | NLM WA 400 | DDC 363.11—dc23
LC record available at https://lccn.loc.gov/2017050773

ISBN: 978-1-138-06866-7 (hbk)
ISBN: 978-1-138-82196-5 (pbk)
ISBN: 978-1-315-74301-1 (ebk)

Typeset in Times New Roman
by Apex CoVantage, LLC
Printed and bound by CPI Group (UK) Ltd, Croydon, CR0 4YY

Contents

Figures

Tables

Annexures

Foreword

I was asked early on, when this book was being drafted, to consider writing a fore-word and so I have been fortunate to have read the early drafts of this book. This has allowed me to reflect on the impact, not just that the MBA had on my career, but also, now that I am in a senior leadership position, just how relevant the skills were that I acquired as a safety practitioner, to my roles in management and leadership. I have concluded that many of the skills are transferrable, and actually the approach you take as a safety and health practitioner is directly relevant to your role as a leader.

Strategic leadership is usually about making choices to lead you to an imagined future. Evaluating alternative strategies requires considering the costs and benefits of each against the risk. That is a very similar process to making a risk assessment and devising risk controls that are proportionate to that risk. The additional consideration for the leader of an organisation is of course foreseeing the impact of your decisions on every part of the organisation as well as your stakeholders, customers, employees and market place. This wider critical thinking and evaluating impacts from a number of perspectives is a key theme in the first chapter of this book.

When leading change or implementing strategy, the goals of every part of the organisation need to align behind your chosen strategy, and also the structure, the people, the resources, the processes and the culture. Implementing any programme of safety management requires all of those elements to make it happen.

Risk management is an essential part of my role as a chief executive. Ensuring that we have systematically identified and evaluated the wider risks to the organisation, put in place controls which address the root causes, and mitigation should the worst happen, including business recovery plans with our key suppliers is essential to ensure that I sleep at night. The sections of this book dealing with enterprise risk management are enlightening in this respect.

For any organisation or function within a business there will be "critical success factors", things which you absolutely should get right. Setting your key performance indicators – those measures that tell you whether you are consistently and reliably meeting your critical success factors, and give you an early warning if you are drifting to failure, is a concept which many practitioners will recognise as akin to leading and lagging indicators of performance.

When things do go wrong, root cause analysis – the skill all practitioners need in accident investigation – is a tool that I have coached the non-safety practitioners in my organisation on. Analysing a problem, and researching and devising a solution that deals with the cause and not the symptom, is a process transferable from safety and health to managing any issue.

As a safety and health practitioner, much of my day to day activity was negotiation, influencing and persuading others to integrate safety into their activities and decisions. Those skills are still called on whether it is debating priorities with my direct reports, or making a case for investment in innovation to our Board.

My own MBA, completed in the late 1990's, had a profound impact on my career. It was a programme with experiential and reflective learning at its core, and was the inspiration for the book that Waddah and I published in 2014 on Reflective Learning for the Safety Practitioner. Hence, I was very excited about Waddah and Rob's project to reflect and apply their learning on their MBA programmes to the work of the safety and health practitioner.

Development is never easy. Self-development is perhaps even more difficult. This book links familiar concepts explored in studying for the NEBOSH Diploma and within the substance of the new ISO 45001 standard on safety management systems, and develops them into a senior leadership context. I particularly commend the model that Waddah and Rob have introduced to give an overview of the framework for transformation which is at the heart of this book. One key skill stressed to me on my MBA was the need to have a "helicopter view." In other words, to be able to look at the organisation in its context, its economic and market environment, to see its vision, purpose and strategy – the "big" picture. And then to be able to look in more detail at its component parts, its functions and its processes. The first two sections of the book and the model consider these aspects – the "strategic" and the "cross-functional".

The final section of the book and the model intriguingly deals with being "distinctive" – developing your own skills to use data and analysis to make better decisions and to make a convincing case, to innovate in your professional practice and communicate more effectively. Knowing yourself, reflecting on your strengths and using them effectively are crucial skills and indeed habits which will help you to become a better leader and the best safety and health professional that you can be.

It is a source of some frustration to me that often the director responsible for safety on a board is not a qualified safety professional. Why shouldn't safety and health professionals aspire to a seat in the boardroom? Waddah and Rob have produced an accessible but authoritative guide to taking your leadership skills to the next level to break through that glass ceiling.

Teresa Budworth, CEO – NEBOSH
December 2016

Preface

As an Occupational Safety and Health (OSH) practitioner have you ever wondered "How can I shape my career trajectory to reach a C-suite position in business?" Or perhaps – for those who do not aspire to positions of this nature: "How can I develop my ability to persuade and influence top management more effectively?" These questions formed the backbone of the research for this book and are answered within to assist you in achieving your personal and professional OSH goals.

Some years ago, we both made the decision to attend Business School with a view to progress from our current functional roles to broader corporate positions. However, we could never have imagined that the lessons from our MBA studies would have changed our understanding of the positioning of OSH within a business so dramatically. The intention of *The 10 Step MBA for Safety and Health Practitioners* is to share the key lessons from Business School within the specific context of OSH to enable you, the OSH practitioner, to create your own path to power.

Risk is now firmly on the agenda of organisations, with frameworks for Enterprise Risk Management, Business Continuity and Resilience prevalent, as Executives seek assurance that risks are being effectively identified, assessed and controlled. Against this wider backdrop of risk, OSH practitioners increasingly need to have the ability to talk to business units in strategic and commercial terms, with a view to creating solutions that are integrated with wider business considerations. These are the challenges that OSH practitioners are facing, but unfortunately, are not always equipped with the knowledge and capabilities to tackle head on.

We are confident that this book will be a worthwhile investment (and far cheaper than an MBA!) in enhancing your understanding of the business realities of decision making and to differentiate yourself in the competitive OSH job market. So, whether you are an energised OSH professional looking to take your career to the next level, or a disenchanted practitioner looking to be rescued from corporate obscurity . . . read on and find out how you develop influence by becoming more *strategic, cross-functional* and *distinctive* in your approach.

With optimism,

Rob Cooling and Waddah S Ghanem Al Hashmi
October 2017, Dubai, UAE

Introduction

Primary purpose

The purpose of this book is to help Occupational Safety and Health (OSH) practitioners improve their understanding of the language of business to develop influence and bring about change. We all want our organisations to achieve continual improvement in OSH standards, culture and performance, but to promote improvement we need to be equipped with the right business skills and know-how. Business theories, concepts and tools can be acquired from various sources; however, the key value of this book is the presentation of this information within an OSH context.

Key takeaways

The research undertaken in preparing this book identified the need for a transformation in OSH practice, for OSH professionals to position themselves more effectively in the ever-changing world of business. Although the word "transformation" may seem somewhat radical, it is deemed that a substantial change is needed in approach, particularly for those aspiring to reach senior OSH positions.

The book is presented over 10 Steps encompassing a typical MBA programme. Although the 10 Steps do not have to be read sequentially, the book does establish three key themes which are deemed critical in understanding the world of business to exert greater influence.

- **Strategic** – *aligning OSH to the overall direction of a business and creating a lasting OSH purpose that all stakeholders can relate to*
- **Cross-functional** – *understanding the different parts of an organisation and integrating OSH within business functions and ways of working*
- **Distinctive** – *looking for creative new ways of presenting OSH data and information to generate interest and enthusiasm.*

To help clarify the relationship between the 10 Steps within the MBA programme and the three key themes identified above, the Occupational Safety and Health practitioner transformation model is presented in Figure 0.1.

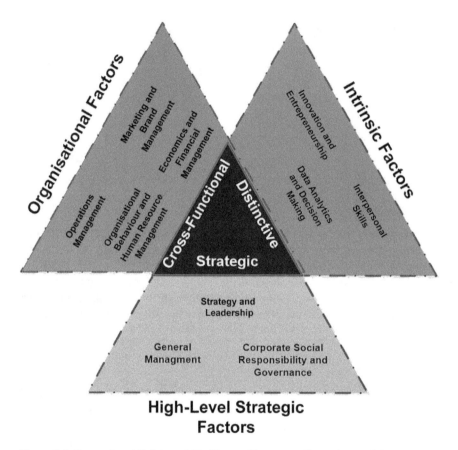

Figure 0.1 Occupational Safety and Health practitioner transformation model

Intended audience

This book is intended for OSH practitioners working in strategic roles, with responsibility for OSH across multiple sites and jurisdictions. However, it is anticipated that those aspiring to reach senior OSH positions will also find the content enlightening, including those considering whether to embark on an MBA programme and assessing its relevance to our profession. Furthermore, top managers and those working in other business functions (e.g. Marketing, Finance, Human Resources, etc.) may also draw lessons from the book, particularly in relation to the strategic and functional aspects of OSH management.

Increasingly, OSH practitioners have responsibilities for environmental management. Although the book is not written within the context of Health, Safety and Environment (HSE), all the Steps of the MBA programme have relevance to environmental management, with the recognition that the language of business is just as important to environmental issues as to OSH matters. Indeed, environmentalists may well find much of the content of this publication useful, having to face similar challenges to OSH professionals in convincing organisations of the added value of environmental protection.

Overview of chapters

Following this introductory discussion, Step 1 of the MBA programme covers *Strategy and Leadership*. The importance of creating an OSH strategy aligned to corporate strategy is considered, including setting an OSH vision, supported by a clear road map for realising the vision. Strategic frameworks are presented, coupled with evaluation on their application to OSH management. The concept of strategic agility is also outlined with insights on how organisations can build resilience into their OSH strategy to respond to significant change. The first Step also addresses the role of leadership with the understanding that leadership is not defined by organisational position. Several key leadership styles are presented, along with discussion on leadership theories and frameworks applicable to OSH.

Step 2 addresses *General Management*. This Step considers the management of an enterprise, providing the OSH practitioner with the essential skills needed to manage risks and opportunities from a holistic perspective. This Step of the program addresses the requirements contained in ISO 45001, with the understanding that General Management within the context of OSH, centres on the development, implementation and maintenance of an OSH management system. Importantly, new terms, concepts and approaches within ISO 45001 are covered, supported by discussion on the step-change in OSH practice required by the introduction of this new standard.

Step 3 provides an analysis of *Corporate Social Responsibility and Governance*. Key terminology is presented with an evaluation of how these concepts can be used as vehicles to promote OSH. The idea of the organisation as a corporate citizen is addressed with discussion on the moral and ethical responsibilities of organisations. The current focus on sustainability is broached and how OSH relates to and supports the sustainability agenda. This Step considers the ongoing shift to risk management raising awareness of enterprise risk management and the opportunities created for OSH practitioners. Best practice corporate OSH governance is also reviewed, with guidance on integrating OSH within governance arrangements.

Organisational Behaviour and Human Resource Management is considered at Step 4. The importance of understanding organisational culture and cultural maturity is discussed focusing on the design and implementation of behavioural change programs in consideration of emotional, rational and situational factors. Human resource management is addressed, including the value of creating a strategic alliance between human resources and OSH functions to leverage cross-over points and collaborative opportunities that exist. The growing focus on well-being is also covered in this Step and the need for OSH practitioners to ensure that occupational health and well-being issues are embraced.

Economics and Financial Management is dealt with during Step 5. Microeconomics principles, frameworks and tools applicable to OSH are explored. Consideration is also given to wider macro-economic issues and the need for the OSH practitioner to be aware of the impact of macro changes on the OSH landscape. This Step presents discussion on the importance of improving financial acumen to develop levels of influence. A proactive approach is encouraged whereby the OSH practitioner applies financial tools, including net present value and cost-benefit analysis, to determine the value added by OSH interventions.

Step 6 addresses *Operations Management*, considering the oversight, design and control of business processes. Discussion is provided on the development of business management systems and the arguments for and against the integration of management systems. Key operations management issues are considered, including procurement, outsourcing and supply chain management and their links to OSH. Relevant theories are also briefly presented, such as Total Quality Management, Six Sigma and lean manufacturing, supported by analysis of their implications to OSH.

Marketing and Brand Management is considered during Step 7. The Step provides insights in relation to how OSH can influence the development, selection and execution of marketing strategies for an organisation's product and/or service offerings. The importance of OSH outcomes to brand management are discussed, with specific examples of how OSH can be embedded within marketing strategy. This Step also outlines how the OSH practitioner can take on the role of marketer to raise the organisational profile of OSH, along with improving their individual status and position in the organisation. Guidance is provided in relation to how the OSH practitioner can recognise their core competencies and build a compelling value proposition and narrative for their role.

Step 8 addresses *Data Analytics and Decision Making*. The Step begins with discussion on the essence of decision making and the importance of objective data and analysis in reaching sound and robust business decisions. A range of analytical tools, including descriptive and inferential statistics, are covered to enable the OSH practitioner to learn how to quantify and present uncertainty and to be able to ask the right questions when presented with OSH data. This Step also considers the role that luck can play in OSH outcomes and whether an environment can be created to increase the propensity for luck.

Step 9 discusses the importance of *Innovation and Entrepreneurship*. The concept of innovation is presented, supported by an innovation typology, providing examples of how innovation can be embedded in OSH practice. Importantly, this Step also discusses change management and how organisational systems, infrastructure and capabilities can be created to encourage innovation and how experimentation can be encouraged within the confines of management control. The OSH practitioner as an entrepreneur is also posited, in recognition of the desire of many OSH professionals to establish their own business.

The final Step in the 10 Step MBA programme is *Interpersonal Skills*. Discussion is presented on the concept of emotional intelligence and its relevance to career progression. Key interpersonal skills including communication, teamwork, conflict resolution, empathy and curiosity are evaluated, providing the OSH practitioner with an understanding of the softer skills needed to generate change in organisations. The Step considers how coupling theoretical knowledge, experience and interpersonal skills can enable the OSH practitioner to succeed within an organisation. The importance of power in organisations is also assessed, supported by examples of how to develop the ability to influence and persuade individuals at different levels of an organisation.

The book closes with *Conclusions* which revisit the overarching purpose of the book and associated objectives. The key themes of the book are consolidated, including the importance of OSH practitioners continuing to look for new ways to be more strategic and cross functional in their approach, along with being distinctive in identifying innovative and new approaches to OSH practice. Reference is also made to the potential next steps for the OSH practitioner to realise their newly discovered path to power.

Method

To reach the conclusions presented in this book, a range of techniques were utilised, including a desk based review of relevant literature. As OSH is a cross-disciplinary subject, consideration was given to literature across a range of different fields, including strategic and organisational management, along with specific OSH studies. The desk-based review was accompanied by a survey conducted by the authors in collaboration with the Institution of Occupational Safety and Health (IOSH), the world's largest OSH professional membership organisation. This survey, referred to herein as the *OSH Practitioner Insight Survey* was supported by survey results provided by the National Examination Board in Occupational Safety and Health (NEBOSH), the world's leading awarding body in OSH management and practice, which was used to inform recent revisions to the NEBOSH Diploma syllabus in November 2015. These surveys were supplemented by a number of focus groups with senior OSH professionals to explore the key themes emerging from the research.

Literature review

A range of different business management and OSH related literature was reviewed in the development of this book. These sources include text books, scholarly articles and legal encyclopaedias. The review of relevant literature focused on academic texts, journal and magazine articles and research reports, particularly from the UK Health and Safety Executive (HSE). The HSE produce numerous codes of practice and guidance notes relating to OSH management, which were analysed throughout the duration of time working on the book. Information was also gathered from a number of internationally recognised organisations within the sphere of OSH, including the International Labour Organisation (ILO) and World Health Organisation (WHO), which helped to ensure that discussion is presented from a global perspective.

OSH practitioner insight survey

Our survey, supported by IOSH, captured perceptions and views regarding OSH within the context of business. The survey incorporated a total of 30 statements linked to the relevance of business education to OSH and asked participants to indicate the extent to which they agreed or disagreed with the statements. The survey was sent to 20,000 IOSH members with 2,039 responses (10.2%). Throughout each Step of the programme references are made to the findings from the survey with the list of statements included in Annexure A.

NEBOSH Diploma consultation

The findings of a survey conducted by NEBOSH as part of a consultation exercise for updating the NEBOSH Diploma qualification also influenced content within the book. Students who studied one of the Diploma variants (NEBOSH National (UK) and International Diplomas) were invited to take part in a survey to determine if changes were required and whether the qualification and its current content was applicable to their job roles. This survey was distributed to 6,500 students with 1,350 responses (21%). The survey identified a range of potential new content, including:

- OSH leadership – role of leaders in setting direction
- Professional practice – influencing, coaching, negotiating with stakeholders
- Corporate social responsibility (CSR) – role of OSH in the CSR agenda
- Budgetary/economic constraints – making the case for OSH expenditure.

Focus groups with OSH professionals

To substantiate the empirical data from the OSH Practitioner Insight Survey, several focus groups were conducted with senior OSH professionals working in different jurisdictions. The focus groups were conducted involving the use of a semi-structured questionnaire drawing on the results on the survey. These sessions provided an opportunity to prompt further discussion and exploration of the key themes explored in the book. Discussion on the different topics throughout the book has been reinforced by the insights gathered from these focus groups and conversations with seasoned OSH practitioners, ensuring that the content is reflective of theory and practice.

The discipline and profession of OSH

Although many people reading this book will have a sound knowledge of OSH management and practice, it is useful to provide some brief introductory context for the uninitiated. OSH is a discipline concerned with the prevention of death, injury and ill health to those at work and those affected by work activities.

Essentially, it is a multi-disciplinary subject reflecting a broad range of areas, each of which has its own discrete research community. However, there are two distinct aspects to this organisational discipline: namely safety *and* health. Occupational safety focuses primarily on accident prevention and reducing the risks associated with activities, which may lead to injury and damage to plant and equipment. Common safety hazards include work at height, machinery and moving vehicles. In comparison, occupational health is concerned with exposure to hazardous agents, such as chemical, physical and biological hazards, which may bring about changes in an individual's health status.

In comparison to established professions such as finance, engineering and construction, OSH is a relatively new and emerging profession. However, OSH can clearly be regarded as a profession as it encompasses a number of defining characteristics, common with other recognised professions:

- Possession of specialised knowledge acquired through completion of a period of study, supported by continuous training
- Presence of professional membership organisation(s) and a defined membership structure
- Codes of conduct or ethical standards to guide behaviours of those in the profession.

In 2002, IOSH was granted Royal Charter and the power to award OSH professionals the title of "Chartered Safety and Health Practitioner". Historically, many individuals had practiced OSH as a part-time component of their job role. However, with the inception of chartered status and the increasing global demand for competent practitioners, OSH has become a growing career option with a plethora of professional, undergraduate and postgraduate training programmes available for aspiring candidates. These developments have helped to develop further recognition of OSH as a profession and ensure that OSH practitioners gain parity in the workplace compared to other business professionals.

The publication of ISO 45001 has further strengthened the importance of OSH as a key strategic prerogative and provides greater opportunities for aspiring OSH practitioners. Annexure B refers to the key clauses within ISO 45001 with cross-referencing to the relevant content in the book. A similar exercise has been performed, detailed in Annexure C, against the November 2015 syllabus of the NEBOSH International Diploma, which is still regarded as the pre-eminent qualification for OSH practitioners looking to reach senior level OSH positions.

Interaction guide

Reading and concentrating is hard work—your brain is consuming glucose, oxygen and blood flow and will soon get tired! To help keep you interested (and reading!) the book is embedded with valuable MBA lessons from theory and practice that can be applied in your role as an OSH practitioner. There are also quick memorable tips, findings from our OSH practitioner insight survey and questions to establish whether you are taking in the key messages. Look out for the following icons as you travel through the 10 Step MBA programme.

 THEORY – frameworks, tools and techniques for use at work.

 PRACTICE – cases studies and examples from around the world.

 TOP TIP – simple advice to increase influence and impact.

 SURVEY – findings from our OSH practitioner insight survey.

 TEST – exercises to determine levels of understanding.

Part I

Strategic

Strategic

Strategy and
Leadership

General
Managment

Corporate Social
Responsibility and
Governance

High-Level Strategic
Factors

1   Strategy and leadership

Welcome to Step 1 of the *10 Step MBA for Safety and Health Practitioners*! We begin with *Strategy and Leadership*, arguably one of the most important lessons in your path to power as an OSH professional. This Step relates to the first of the three themes within the OSH Practitioner Transformation Model on the need to "be strategic". Strategy concepts will alter the way you look at your organisation, providing you with frameworks and tools needed for OSH direction, while leadership skills will ensure that you can implement these approaches in practice. So, let's not delay any further. It's time to start your journey of transformation as an OSH practitioner!

On completion of Step 1, you will be able to:

* Explain what is meant by strategy and the importance of making trade-offs
* Appreciate the importance of analysis and diagnosis to effectively develop a vision and strategy for OSH
* Create a framework and road map for executing an OSH strategy
* Understand how to build and maintain agility within your OSH strategy
* Recognise the links between strategy and leadership and the different types of OSH leadership.

Strategy

What should be the first question you ask when embarking on a journey to enhance OSH performance? The answer is simple – *'What is our OSH strategy?'*. Unfortunately, the focus for many OSH professionals when addressing OSH performance is to begin with the OSH policy and associated management arrangements. Although an OSH policy and management arrangements are important, and in many jurisdictions legal requirements, they will not provide all the answers to addressing the challenge of generating continual improvement in OSH performance—for this you will need a strategy!

What is strategy?

Strategy has always been a vogue term in business. Executives proffer that strategies are needed before rushing head-first into tactical decision making;

however, they sometimes struggle to understand the nature and content of effective strategies. Fundamentally, strategy is about making choices, typically concerning *what* you want to do and *how* this will be achieved. From a wider business perspective, this may include what products and/or services are offered and how business decisions are executed. Strategy also involves making trade-offs as we cannot do everything.

In relation to strategy formulation, it is important to recognise that key choices are made at different levels of an organisation. When developing an OSH strategy an understanding and alignment with strategies existing at these various organisational levels may address:

*   Corporate Strategy – defining organisational purpose and values, decisions on which business to be in, how diversified the organisation should be, where to invest and where to divest, how to allocate capital, how to manage the relationships between the different geographies (where applicable) and business units
*   Business Unit Strategy – considering market positioning, strategic objectives for growth, return on investment, profitability, cash generation and collection
*   Functional Strategy – understanding how functions contribute towards the translation of corporate and business unit strategy, managing resources, establishing performance criteria and performance improvement targets.

 91% of respondents either "agreed", or "strongly agreed" that I am confident that I can think strategically on OSH matters.

All organisations wish to improve OSH performance, but there are many choices to be made in determining what interventions to implement. Our OSH Practitioner Insight Survey indicated that most OSH practitioners believe they can think strategically on OSH matters. So where should you start when it comes to preparing an OSH strategy? Well, many organisations look at what they did last year and replicate the same! Indeed, time pressures and budgetary constraints can make it difficult to allocate sufficient resources for strategy development, but periodically it is important to take a step back and undertake some exploratory work before building your OSH strategy.

 Remember that strategy is about making choices and that the essence of your OSH strategy is often more about the things that you decide not to do.

Diagnosis before development

It is difficult to build your organisation's OSH strategy without some level of diagnosis. You would not like it if you went to the doctors and treatment was prescribed before any tests had been carried out! The established links in research between OSH culture and performance (Health & Safety Executive, 2002) suggest that a good starting point in understanding the OSH priorities and barriers within your organisation is to undertake a survey of OSH cultural maturity.

It is important to determine the maturity of your organisation's existing OSH culture, as this can represent a key determinant in ensuring the successful implementation of OSH interventions. A number of excellent proprietary models are available to assist in the process of evaluating OSH culture, including the Health & Safety Laboratory's (HSL) Safety Climate Tool (Health & Safety Laboratory, n.d.). This tool includes a range of statements related to the OSH climate of your organisation to which participants are required to confirm the extent to which they agree or disagree with the statements. The outcome is unique insights into perceptions of your OSH culture supported by suggestions for improvement.

Although these types of tools can be useful in determining the pillars of your OSH strategy, it is important that the following points are considered:

- Tailor questions to address industry/organisational specific requirements
- Acquire a representative demographic sample, including different levels of the organisation (e.g. senior management, middle management, frontline workers, etc.) and make-up of the labour force (e.g. employees, contract labour, agency workers, etc.)
- Provide the survey in alternative languages with translation support for workers who do not speak English as a first language
- Use alternative forms of survey delivery to ensure a high completion rate (e.g. email link, face to face sessions, etc.)
- Follow up with interviews and other focused study groups to determine high priorities and barriers.

The growth of on-line survey tools, such as Survey Monkey (Survey Monkey, n.d.), also means that options are available for creating your own tailored surveys, at a much lower cost. So, with a little thought and effort you can undertake your own studies of the prevailing OSH culture to understand what to do next. When information relating to your OSH culture is coupled with OSH performance data and consideration of OSH risks and opportunities you will be in possession of all the building blocks necessary to define your OSH vision.

Building your organisation's OSH vision

Once you have completed a diagnosis of the existing situation, the next step in creating an effective OSH strategy invariably involves defining a vision. Without the need to name and shame organisations, a quick trawl through any search engine will quickly find proclamations of visions for 'world class OSH performance'; but a vision should be something that you can see – hence the term vision!

Furthermore, you should know when the vision has been achieved and it should reach out to everyone within the organisation at a personal level to influence desired behaviours. The ability to describe the future you desire in terms that employees can understand is a crucially important part of defining an OSH vision. It helps people visualise the journey you want them to take and ensure the decisions they make are consistent with that vision.

The term 'zero' is commonly adopted into OSH visions – zero harm, target zero and beyond zero. Indeed, a vision that incorporates zero can be a powerful and aspirational goal; however, take time to consider the context of your organisation to determine whether the attainment of zero harm is feasible. Due to concerns regarding the unintended consequences of zero based visions, many organisations are now moving towards more positive and value based OSH visions, driven by progressive thinking, such as "safety differently", coined by Sydney Dekker (2014). These new approaches centre on the belief that people are the solution and not the problem when it comes to driving improvements in OSH performance (see Step 4 on *Organisational Behaviour and Human Resource Management* for further discussion on moving from a performance based organisation to a learning organisation).

 Don't fall into the trap of a 'copy and paste' zero harm vision. Take time to understand OSH priorities and barriers and create a vision appropriate for your business.

Once a vision has been formalised it needs to be communicated effectively to the workforce. Training provides an ideal opportunity to communicate the vision; however, it should be reinforced periodically using other forms of communication, such as posters, leaflets and text messages, always with involvement from top management, to ensure that the critical messages remain in the minds of the workforce. These initiatives will help in developing employee commitment and ownership of the OSH vision and will increase the chances that the vision will be achieved in practice.

Developing a strategic road map

An effective OSH strategy should provide clear direction in terms of the activities required to achieve your OSH goals; however, the creation of a glossy and compelling vision is not enough – you need to ensure that this is supported by a clear

Innovation in OSH strategy

In 2015, Laing O'Rourke, one of the UK's largest contractors, starting trialling an innovative new OSH strategy. The organisation had become sceptical of the 'zero' vision for OSH and decided that a different approach was needed and in June 2015 formally rolled out the new strategy across the Australia business.

Central to the OSH strategy are three key principles:

1 People are the solution, not the problem – people putting processes into practice must not only have the competence, but the confidence to make the right decisions.
2 Safety is the presence of positives, not the absence of negatives – bright spots should be identified and replicated across the organisation.
3 Safety is an ethical responsibility, not a bureaucratic activity – safety management is about protecting people from harm, not protecting companies from litigation.

Laing O'Rourke's new OSH strategy is a great example of a willingness to embrace change and a new direction when analysis indicated the current approach had stalled.

It is important to have the courage to make changes when current strategies are not working, so make sure you don't fall into the trap of letting OSH strategy stagnate.

implementation plan that lays out the journey required to achieve your vision, detailing associated actions, priorities, time-frames and responsibilities.

Achieving an aspirational vision may be regarded as unrealistic by the workforce, particularly if the organisation is currently suffering from poor levels of OSH performance. In this type of situation, change needs to be broken down into more manageable phases to instil the confidence among the workforce that OSH goals are realistically attainable. Presenting a vision as a journey, as opposed to a short-term requirement may help in developing the mind-set that gradual improvements in OSH performance may ultimately lead to accomplishing the goal.

In line with this approach, a road-map can be created, incorporating sub-goals. These more concrete and proximal sub-goals may help prompt initial movement towards the ultimate destination. Often the most difficult aspect of any change effort is initiating people to change their behaviour, but if you shrink the change, all of a sudden, the goal is perceived as more visible and within reach (Heath & Heath, 2010).

When sub-goals are created, it is important to measure OSH performance against these goals periodically. However, there often can be a behavioural tendency to focus on negative aspects when assessing performance. From a behavioural perspective, it is also critical to emphasise how far we have come, as opposed to how far we have to go. This will help to create a sense of accomplishment and generate the impetus that is needed to improve further (Step 4 on *Organisational Behaviour and Human Resource Management* provides more information on the design and implementation of behaviour change programs, which can play an important role in improving OSH performance).

OSH strategic frameworks

An important component in developing an OSH strategy is to establish strategic outcomes or objectives, each one deemed necessary to achieve your overarching OSH vision and mission. Associated programmes (or projects) will also need to be established to ensure that each strategic objective is achieved in practice.

Try applying this model when creating your own strategic OSH framework. Also make sure to create an implementation plan incorporating milestones for completing each programme, remembering that effective programmes should detail the responsibilities, resources and timeframes for completion.

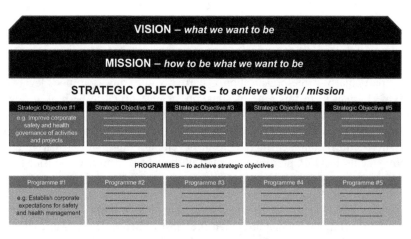

Figure 1.1 Strategic framework for Occupational Safety and Health

Building strategic agility

So, you have developed a comprehensive OSH strategy that paints a clear picture of what the future will look like – now all you have to do is push on and implement the actions in your strategy and realise your vision – well, it is not quite as simple as that! The reality is that even with the most well-crafted strategy it is unlikely that you will achieve all your OSH goals, as over time things change, or as Henry Mintzberg (2013) would say – stuff happens! We live in turbulent times with a wide range of internal and external variables that can impact OSH standards and ultimately performance. Subsequently, it is important to build a level of agility into your OSH strategy for your organisation to be resilient and responsive to significant change.

The introduction of ISO 45001 has increased the emphasis on OSH strategy and leadership. Significantly, the standard follows the requirements of the high-level structure set out in Annex SL (ISO, 2015), which is used as a framework for other ISO management standards, including ISO 9001 and ISO 14001. Annex SL includes a few new ideas, including 'context of the organisation'. This concept requires you to examine the internal and external factors which could affect your business, such as changes in employee relations, new materials or technologies, and then look at the risks and opportunities they present (see Step 2 on *General Management* for more information on internal and external influences). If we can understand uncertainties better then we have an improved chance of responding to them and improving our strategic agility.

To identify emerging OSH issues, organisations should establish capability for horizon scanning. The Health & Safety Executive (HSE) undertakes horizon scanning and other future activities to increase awareness of developments, trends and other changes in the world of work (Health & Safety Executive, 2013a). This includes building a range of possible futures to identify and describe their OSH implications, with current issues, including energy topics, political, science and technology, socioeconomic and the workplace (Health & Safety Executive, 2013b). Although scenario building may not be necessary for most organisations, an understanding of horizon scanning can be important in developing resilience and enabling your organisation to become more pro-active in identifying potential changes which may impact OSH standards.

 Prepare a list of potential OSH risks and opportunities facing your organisation in the coming years. Prioritise these findings and consider how you intend to mitigate risks and realise opportunities.

Stakeholder management is another critical component in creating an agile OSH strategy. To understand the landscape in which your organisation is operating, it

Stakeholder management

Stakeholder management has long been recognised as a pillar of effective project management, but as an OSH practitioner have you identified all the stakeholders in your sector and determined their relative importance in achieving your OSH performance goals? Well, if not, why not start today!

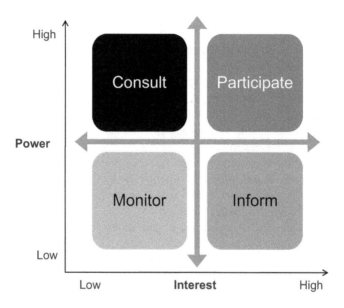

Figure 1.2 Stakeholder management matrix for Occupational Safety and Health

Make a list of all the stakeholders in your sector and then draw a simple two-by-two matrix (MBAs love two-by-two matrices!). On one axis is the level of *Interest* that the stakeholder has in your OSH performance. For example, an OSH regulator may have a high level of interest in how many incidents you have. On the other axis is the level of *Power* that the stakeholder has to influence your OSH performance, for example, if you have a main contractor on your site that employs many people then they may have a significant impact on your OSH performance. The position of each stakeholder on the matrix will dictate your intervention strategy. Notably, if you have stakeholders that have both high level of Interest *and* Power in terms of your OSH performance, then it is critical to employ a participative approach, encouraging their active involvement in your OSH decision making.

is fundamental to identify the stakeholders capable of influencing your organisation's OSH performance. The reality is that achieving good OSH performance is a partnership approach, but many organisations do not fully consider the implications of actions from others outside of their organisation and managing related interfaces. The concept of 'context of the organisation', within ISO 45001, requires organisations to spend more time considering the roles of stakeholders, particularly when evaluating external influences on OSH.

Leadership

So now that you have developed your OSH strategy – what now? Well, strategy is nothing without execution – and for that you will need leadership! The first misconception to dispel surrounding leadership is that it has nothing to do with your position in the organisation – everyone is a leader! The ability to bring about a transformation in business practice is not solely restricted to top management. Therefore, it is important that you identify your leadership style and learn how to employ it effectively.

Interesting research on leadership suggests that there are two broad forms of leaders (Bass, 2008):

1   Task orientated – focused on goals, objectives and targets and getting things done
2   Relationship orientated – concerned with communication, consultation and fostering good relationships.

If you perform a quick self-evaluation you will probably determine that you are weighted towards either being a task orientated or a relationship orientated leader. Importantly, one is not superior to the other; you will need different leadership styles at various times in your career to address changing situations. There may be aspects of your leadership style that you believe undermine your chances for success, but don't worry too much about this; try to identify what type of leader you are and build on the positive aspects of your personality and ensure that they are reflected in your leadership style.

Types of OSH leadership

Research has indicated that there are a range of different types of OSH leadership, as detailed below:

*   Transformational – this form of leadership is characterised by the creation of positive change through charisma, inspiration, individual consideration and intellectual stimulation
*   Transactional – this approach is focused on the day-to-day transactions involved in running an organisation and ensuring tasks are completed

- Servant – a servant leader is someone who leads by meeting the needs of the team. This term typically relates to a highly supportive form of leadership, although the individual may not necessarily receive formal recognition as a leader.

Situational and contextual leadership

An understanding of different types of OSH leadership is important, however, your effectiveness as a leader will be judged by your ability to adopt a leadership style reflective of the situation you are presented with and the decisions that must be made. As indicated previously, different approaches to OSH leadership are needed at different times, but how do you know which is the best approach to adopt? Well, often this will come down to your gut instinct; however, research has established a more objective and contextual approach to help arrive at the most appropriate decision-making process as a leader.

The Vroom-Yetton-Jago Decision Making Model of Leadership (Vroom et al., 1988) presents a range of leadership styles ranging from participative to highly autocratic with a range of factors relating to the decision, such as quality, commitment from group members and time restraints to determine the most appropriate style of decision making. Although this method presents an interesting mechanical procedure to arrive at a decision-making process, discussions within Step 10 on *Interpersonal Skills* argue that your softer skills will arguably be your strongest ally as a leader, ensuring that your decisions are in tune with the emotional and often political subtleties within your organisation.

Leader-member exchange theory is another leadership theory that has relevance to OSH management (Graen & Uhl-Bien, 1995). Leader-member exchange identifies "in-groups" within organisations, who typically assume greater responsibilities, rewards and more participation in decision making, and "out-groups", who may have less access to managers and do not receive the same opportunities for growth and advancement. People within the out-group may be more inclined to exhibit unsafe behaviours, largely due to feelings of detachment and a lack of motivation. Subsequently, efforts should be made to increase the size of in-groups to increase the propensity for safe behaviours. As an effective OSH leader, you can contribute to the creation of in-groups by treating people fairly and judging on performance not personality.

Sometimes we listen to important business or political officials in the media and feel that we want to mimic their leadership style. Perhaps you have been impressed by leaders like Jack Welch, Richard Branson, or maybe Barack Obama and thought, I want to be like that! Well, remember that personality predicates leadership and that you cannot be someone who you are not. One of the most important aspects of being an effective leader is to be authentic. People soon see through a phony, so stay true to your personality.

Strategy and leadership are synonymous. You cannot be a great leader without being a great strategist and vice versa. One of the aspects of strategy identified earlier was the importance of making choices. Indeed, your effectiveness as a leader will be founded on the choices that you make. As a leader, you have an obligation to bring about improvements in your organisation and that will require change.

Your ability to identify the need for change and more importantly to execute the required actions will result in changes in the way you are perceived at work.

Remember that execution of your OSH strategy is critical. Without execution you will not attain improvements in OSH culture, standards and performance.

In many respects, the implementation of your OSH strategy will be the most difficult and many organisations fail when it comes to execution of strategy. When the going gets tough and you start receiving pushback from different parts of the business, stand true to the choices that you have made and make sure to implement them in practice. Don't be afraid to rock the boat, as strategy will not succeed in a void and your leadership skills will make the difference between rhetoric with no impact and concrete changes which have a tangible impact on your OSH performance.

Defining an OSH vision

So, what makes a good vision? Research has suggested that there are several key features of a compelling OSH vision (Collins & Porras, 1996). When evaluating an OSH vision, it is important to ask the following questions – does the vision:

- create a vivid image in people's heads that provokes emotion and excitement?
- provide clear direction that people can understand and act upon?
- present an audacious and aspirational time-bound goal?
- establish a picture of the future state of the organisation?

Now evaluate the OSH vision provided below from a leading mining organisation across the four dimensions detailed above.

'. . . *the multinational mining corporation state that their vision is: that, together with our employees, we will create an injury and illness-free workplace where everyone goes home safe and healthy each day of their working life.*'

So, what do you think? Spend some time evaluating the vision in line with the criteria above and identify a few improvement areas.

Summary

Strategy and leadership go hand in hand, hence the reason they are both addressed in this opening Step of the programme. This first Step has identified that strategy relates to choices about *what* you do and *how* this will be done. The *how* is a critical consideration as all strategies need to be executed. Time constraints and other pressures can often result in an approach to strategy creation where previous strategies are re-hashed, but try to ensure that your OSH strategy is reflective of the OSH culture, risks and opportunities within your organisation. Inevitably, this will require time to reach out to people to understand the central issues and perceptions towards OSH.

With respect to leadership, remember that leadership is one of the most important factors in achieving positive OSH outcomes. The way teams are led on OSH significantly influences OSH performance, as the real causes of incidents are often traced back to the decisions of leaders. Furthermore, the beliefs, values and behaviours of leaders determine the OSH culture of our organisation with leadership behaviours setting the tone and dictating the true priority of OSH within our organisation. In relation to your OSH leadership style, be aware of the various types of leadership but remember one of the most important aspects of leadership is authenticity. You can have your heroes in life, but you are who you are – don't try to be somebody different!

Key takeaways

- Strategy is about making choices, typically concerning what you want to do and how this will be achieved – *do you have an OSH strategy and compelling OSH vision?*
- Diagnosis of both corporate strategy and organisational OSH culture should be undertaken before creating an OSH strategy – *have you undertaken sufficient diagnostic work to understand the priorities and barriers to be addressed in your OSH strategy?*
- A road-map is required to detail the journey for achieving your OSH strategy – *is your OSH strategy supported by a clear road map, incorporating sub-goals?*
- It is important to build a level of agility into your OSH strategy for your organisation to be resilient and responsive to significant change – *have you sufficiently explored internal and external factors to build a level of agility into your OSH strategy?*
- Effective leadership skills are needed to execute OSH strategy – *have you taken time to evaluate your leadership style and identified the weaknesses to be managed and the strengths to be accentuated?*

References

Bass, Bernard. (2008). *The bass handbook of leadership: Theory, research, and managerial applications.* New York, NY: The Free Press.
Collins, J., & Porras, J. (1996). Building your company's vision. *Harvard Business Review.* September–October Issue.

Dekker, S. (2014). *Safety differently: Human factors for a new era*. Boca Raton, FL: CRC Press, 2nd edition.

Graen, G. B., & Uhl-Bien, M. (1995). The relationship-based approach to leadership: Development of LMX theory of leadership over 25 years: Applying a multi-level, multi-domain perspective. *Leadership Quarterly*, 6(2), 219–247.

Health & Safety Executive (HSE). (2002). *Safety culture: A review of the literature*. Sheffield, UK: HSE Books, HSL/2002/25.

Health & Safety Executive (HSE). (2013a). *Foresight centre – emerging health and safety issues*. Retrieved from www.hse.gov.uk/horizons/.

Health & Safety Executive (HSE). (2013b). *Horizon scanning – current issues*. Retrieved from www.hse.gov.uk/horizons/current-issues/index.htm

Health & Safety Laboratory (HSL). (n.d.). *The safety climate tool*. Retrieved from www.hsl.gov.uk/products/safety-climate-tool.

Heath, C., & Heath, D. (2010). *Switch: How to change things when change is hard*. New York, NY: Broadway Books.

International Organization for Standardization (IOS). (2015). *ISO/IEC Directives, Part 1 consolidated ISO supplement – procedures specific to ISO*. 6th edition. Geneva: IOS.

Laing O Rourke. (2015). *EEJ: Inside an engineering. Safety – time for a new direction*. Retrieved from www.laingorourke.com/media/eej-2016/safety-time-for-a-new-direction.aspx

Mintzberg, Henry. (2013). *The rise and fall of strategic planning*. New York, NY: The Free Press.

Survey Monkey. (n.d.). Survey Monkey. Retrieved from www.surveymonkey.com.

Vroom, V., Jago, H., & Arthur, G. (1988). *The new leadership: Managing participation in organizations*. Englewood Cliffs, NJ: Prentice-Hall.

2 General management

Now that you understand the fundamentals of strategy and leadership, in Step 2 we move on to discuss *General Management*, which is concerned with the leadership and management of an organisation. During this second Step, we will cover the principles of management within the context of OSH and discuss the qualities of effective general managers and the distinction between leadership and management. Importantly, this Step also addresses the day to day aspects of running a business, including creating and implementing policies and processes and their importance in achieving positive OSH outcomes.

On completion of Step 2, you will be able to:

- Understand the principles of General Management and how it relates to OSH
- Explain the differences between leadership and management
- Recognise the pillars of an effective and efficient OSH management system
- Understand the importance of OSH planning with a focus on the management of OSH risks and opportunities
- Prepare your organisation for the transition to ISO 45001 by raising awareness of new terms, concepts and approaches.

Understanding general management

The term General Manager can mean different things in different parts of the world; however, in the context of this discussion, it refers to someone who holds responsibility for the management of a business, or large function or division within a business. Furthermore, the context also infers the delivery of products or services designed to make the business a profit, as opposed to organisations that

 31% of respondents either "disagreed", or "strongly disagreed" that the training and experience they have received has prepared them sufficiently to be the next General Manager of their organisation.

may be established for charitable purposes. That being said, many of the lessons herein on General Management would also apply to non-profit organisations.

One of the questions asked during the *OSH Practitioner Insight Survey* was whether participants felt that their previous education and training had prepared them to be the next General Manager in their organisation. Interestingly, 31% of respondents either "disagreed", or "strongly disagreed" that they were ready for that transition, representing a high response considered to other questions in the survey. So, what can you do as an OSH professional to equip yourself with the necessary competencies needed for General Management?

Many of the skills needed to be an effective General Manager are addressed in detail in the coming Steps of the programme. However, it is worth noting that a great deal of evidence suggests that those with a commercial or financial background tend to end up in General Management positions. For example, just over half of Fortune 100 CEOs in the US have degrees in business, economics or accounting (Stadler, 2015). Therefore, a dilemma is presented in terms of how you can strengthen your understanding of financial matters if you are not involved in the financial aspects of running a business. Well, the short and simple answer to that is: get involved! As discussed in Step 5 on *Economics and Financial Management*, the absence of profit and loss responsibility does not prevent you from taking an interest in the financial metrics that your organisation uses to measure success.

 Leadership vs. management

During Step 1 of the MBA programme on *Strategy and Leadership* the importance of effective leadership in developing a positive OSH culture was discussed. Now the focus shifts to management as opposed to leadership. Although OSH leadership and management are clearly interrelated, it is important to recognise that they do have their differences. The HSE has produced useful guidance on the key distinctions between OSH leadership and management (Health & Safety Executive, 2015), with key differences presented in Table 2.1.

Leadership is about producing change, whereas management is typically more concerned with ensuring business functions operate efficiently and effectively. The role of a leader involves setting the direction of a company and developing fresh approaches to problems. Managers seek order and control and the rapid resolution of problems, whereas leaders may prefer disorder and a lack of structure. Although this may appear paradoxical, if management controls within an organisation are too rigid it can be difficult to encourage free-thinking, innovation and change.

Try to remember the key distinctions and commonalities when you are looking at OSH leadership and management in your organisation.

Table 2.1 Leadership vs. management in Occupational Safety and Health

Leadership	Management
Creates and communicates a vision for the future	Develops a plan and allocates resources
Encourages others to commit to the vision	Sets objectives and organises a schedule
Motivates and inspires workers to overcome barriers	Monitors situations
Encourages innovation	
Helps the organisation to develop by adapting to changing circumstances	Focuses on order and efficiency
	Ensures standards are met

To be an effective General Manager you also should be good at running projects, a fact no different in the world of OSH. Common examples of activities that you may get involved in that would benefit from a project management approach could include:

- Developing and delivering a programme of OSH training to employees
- Preparing and implementing a framework of new OSH procedures within a management system
- Designing and implementing an intervention to change behaviours
- Deploying a survey to gather employee perceptions on OSH.

Our discussions with many OSH practitioners indicate that they are regularly delivering projects of this nature, however, in many cases have never received any formal project management training. There are a number of important steps that can be followed next time you are assigned an OSH project to ensure that it is successful.

1   Prepare a Project Implementation Plan (PIP). The creation of a PIP is a useful starting point for any project. The PIP should include information pertaining to the purpose and objectives of the project and the specific tasks (deliverables) that will be provided. It is also good practice to include an organisation chart and schedule detailing the timeframes for completing each task.
2   Produce periodic reports on progress for management. It is important to hold periodic meetings to assess project progress and to identify any threats and opportunities. These meetings should be followed by concise reports to management. Remember that management tends to prefer brief reports and dashboards so try to keep it short and simple.
3   Evaluate lessons learned at the end of the project. When you finish a project, it is easy to be sucked quickly into other activities. Find time to reflect on lessons learned and record these into a register for sharing with others who may get involved in similar projects.

Reflective practice is an important skill for OSH practitioners. When you finish a project or a significant piece of work, try to schedule some time in your diary to reflect on the activity and to capture the lessons learned.

General management and OSH

The day-to-day aspects of general management within the context of OSH typically centre on the development, implementation and maintenance of an OSH policy and associated OSH management arrangements. Invariably, these arrangements will comprise an OSH management system that may be an integral component of a wider Business Management System (BMS) capturing the different ways of working within your organisation (see Step 6 on *Operations Management* for further discussion on business process management).

The learning that many OSH practitioners are provided on management systems provides a useful foundation when discussing wider aspects of business management and the effective control of activities. Fundamentally, it is critical that the key parts of a business are managed in a structured fashion. In this sense, international management system specifications for OSH provide a universal technical language that can be applied in any part of the world and help in ensuring order and consistency in OSH practice across different jurisdictions and operations.

OSH management and ISO 45001

You have not bought this book to be told about the "Plan, Do, Check, Act" cycle, or for a generic overview of OSH management systems! However, the principles of effective OSH management do correlate positively with effective general management, in terms of the importance of adopting a systematic approach and establishing a robust framework for managing business risks.

The timing of this book is pivotal with the recent publication of ISO 45001, the international standard for Occupational Safety and Health (ISO, 2018). The following sections will provide an overview of the new structure within ISO 45001 supported by discussion on new concepts and approaches within the Standard, which will inevitably bring about fundamental changes in the current approach to OSH management and broader lessons in relation to general management.

Context of the organisation

ISO 45001 commences with a requirement for organisations to demonstrate a broad understanding of the context in which organisations operate, to ensure your OSH management system responds to this environment to meet its intended outcomes. As discussed in Step 1 on *Strategy and Leadership*, this requires understanding internal and external influences. A useful tool for assessing the environment in which an organisation operates is PESTLE analysis (Berry, 2015).

- Political – differing political systems and factors can influence organisational decision making. This can occur at a national level, where political pressures can impact the priority given by organisations to OSH, or internally, where internal politics may impact choices and ultimately OSH performance.
- Economic – the nature of competition and the market faced by an organisation can influence OSH standards. Furthermore, recessionary pressures can impact the availability of financial resources within the economy and at an organisational level.
- Social – demographic changes and trends in the ways people live, work and think can impact OSH standards. Social expectations can also be a strong influence in improving OSH standards, particularly in response to incidents that have generated dissatisfaction and public outcry.
- Technological – developments in technology can have a profound impact on OSH standards. In some cases, technology provides the opportunity to accelerate improvements in OSH performance, however, there may also be risks associated with new technology which must be managed.
- Legal – legal systems differ around the world and will influence the approach taken by organisations with respect to OSH policy and practice. Furthermore, the scope and extent of OSH regulation and enforcement can also impact attitudes towards OSH and the attention given to ensuring compliance with OSH requirements.
- Environment – environmental and ecological factors also need to be considered by organisations. There can often be a close overlap between OSH and environment requirements, so it is important to consider any interactions that exist and the related OSH implications.

Although originally designed to focus on external factors, the points above indicate that PESTLE analysis is a methodology for assessing internal and external issues. If used properly, PESTLE can be a useful tool in understanding market and business position and the related OSH risks and opportunities.

Leadership and worker participation

ISO 45001 will drive a step-change in the approach that many organisations adopt for OSH leadership. Great leadership is not only synonymous with great OSH performance, but also great business performance. The focus on requirements for leadership in ISO 45001 is linked to a growing recognition that OSH performance comes from the top (Health & Safety Executive, n.d.). ISO 45001 requires leaders to get actively involved in OSH, with the following examples of OSH leadership which can be generated in practice:

- Position descriptions for top management should include OSH responsibilities, focusing on OSH strategy/policy setting and implementation

- A nominated Director can be appointed as an OSH champion to help support the implementation of OSH management arrangements
- OSH contributions should be considered when deciding senior management appointments
- OSH performance should be an integral feature of remuneration for top management.

Management commitment for OSH should continue to be evidenced though the authorisation of an OSH policy, however, from a strategic perspective, it is important to recognise that there may be other key documents that sit above the OSH policy as critical components of effective OSH strategy and leadership, such as a compelling OSH vision (see Step 1 on *Strategy and Leadership* for further discussion on these topics).

To establish clear strategic direction for OSH, ISO 45001 requires a member of top management to be identified as accountable for the OSH policy and the OSH management system. This individual should lead on ensuring OSH responsibilities are properly discharged and be responsible for ensuring that sufficient resources are allocated for OSH developments. This requirement is a critical component of effective corporate OSH governance, discussed in more detail in Step 3 on *Corporate Social Responsibility, Accountability and Governance*.

Defining OSH leadership behaviours

With the increased focus on leadership in ISO 45001, now is a great time to be thinking more about OSH leadership. However, one important question is: what does great OSH leadership look like in practice?

As an OSH Consultant or Advisor, you have been approached by an organisation looking for guidance on ways to define OSH leadership behaviours. There is a great deal of generic guidance; however, this organisation is looking for a framework specific to their organisation. How would you approach this challenge?

So, what did you write down? Well, it would be useful to start with an evaluation of the prevailing OSH climate. This would help to identify employee perceptions regarding OSH leadership, the weaknesses to be addressed and the strengths to be accentuated.

Furthermore, why not ask top management in the organisation what they believe to be the qualities of great leaders. In many cases, there will be commonalities between the characteristics of great leadership and great OSH leadership. This way you will be able to arrive at an inventory of OSH leadership behaviours that are bred within the organisation, which will ultimately assist with ownership and implementation of any proposed approach.

Planning

Organisations continue to operate in environments of uncertainty. Against this background of change, planning is a key activity to try to manage uncertainty, in consideration of internal and external factors. As an OSH practitioner, it is important to ensure that planning receives sufficient attention to provide assurance that the OSH management system can achieve its intended outcome, prevent, or reduce, undesired effects and achieve continual improvement.

Central to OSH planning is the management of risk. One of the key changes with the introduction of ISO 45001 is the requirement to manage opportunity in addition to risk. For many years, enterprise, portfolio, program and project risk management have adopted a broader definition of risk relating to anything that can create uncertainty in the realisation of objectives. In this sense, risk can be considered across two dimensions, namely opportunity and threat, which can have a positive or negative impact on the achievement of objectives, respectively (Ward & Chapman, 2011).

When assessing risks and opportunities, it is important to adopt a lifecycle perspective. OSH risks and opportunities should be evaluated at each stage of the product or service lifecycle, for example acquisition of raw materials, design, production, transportation/delivery, use, end of life treatment and final disposal. Consideration should be given to OSH requirements at the design stage and during procurement, as well as providing information about potential significant OSH risks and opportunities associated with products and services. Indeed, the consideration of risks and opportunities through the value chain of a business reflects one of the key themes within this book, namely the need for OSH practitioners to become more cross-functional in their approach to OSH management. See Step 6 of the program on *Operations Management* for further discussion on these topics.

The introduction of opportunity management within OSH management may cause a level of anxiety for some OSH practitioners. The reason for this is that for many people OSH risk typically conjures negative connotations. If you were

---

Opportunity management

A key development within ISO 45001 will be the requirement to manage OSH opportunity in addition to risks.

Try to identify a range of OSH opportunities that could have a positive impact on the realisation of OSH objectives. For example, if one of your OSH objectives is to improve relationships with external stakeholders, what opportunity could have a positive effect on achieving this objective?

A few examples of potential opportunities are provided in Table 2.2, but try to apply this within the context of your own organisation and provide examples that specifically relate to your activities.

asked to identify an OSH risk, the mind may jump to risks such as working at height or moving vehicles. However, the requirement for opportunity management will drive OSH practitioners to look at risks in a new way, with a greater focus on the positives.

Support

The effective implementation of an OSH management system requires the allocation of resources. It is important to remember that OSH is no different than any other organisational function in that it is battling for resources, substantiating the need for presenting a sound business case for investing in OSH. Indeed, one of the key challenges for OSH practitioners is demonstrating the value added by OSH activities.

A key component in supporting the implementation of OSH management arrangements in ISO 45001 is competency and awareness. Competency is more than just training; it is a combination of factors, including knowledge, ability, training, experience and interpersonal skills. It is important for organisations to ensure the competence of individuals that could affect OSH performance. Some organisations have introduced competency management frameworks which detail job roles or families, associated OSH competency levels required for differing positions and the training and development programs to be initiated to meet competency requirements.

With respect to awareness, there is certain OSH information that all employees should be made aware of, including OSH policy and arrangements and details on OSH risks and associated controls that are relevant to them. The provision of information is one way to help ensure that there is regular dialogue on OSH within organisations and it should be remembered that continual efforts are required to ensure that awareness levels are maintained, particularly following significant change.

The provision of information is a key component within ISO 45001 and can relate to both internal and external sources. Internal sources include the information and data that is produced from inside the organisation, including risk assessment reports, accident records and inspection reports. External sources relate to information that is available outside of the organisation. This type of information may include health and safety legislation, HSE publications such as Approved Codes of Practices and industry specific guidance.

Table 2.2 Opportunities in Occupational Safety and Health

Opportunities
Improving execution of OSH strategy across business functions
Enhancing relationships with stakeholders
Raising brand awareness through positive media campaigns
Improving employee morale through fewer incidents
Enhancing productivity through reduced downtime

If you have workers with low English language skills then you may need support from translators and interpretors to ensure key OSH messages are understood.

Another critical consideration in ISO 45001 to support the implementation of OSH management arrangements is communication and participation. The involvement of workers is widely believed to be central to the achievement of improved organisational performance across many dimensions, including that of OSH management. Subsequently, it is important that organisations look to involve workers in OSH management from the provision of information to more active forms of involvement, such as consultation and participation. Essentially, it is this active involvement of workers that helps to develop loyalty and commitment to the OSH vision and will help to ensure that OSH goals become a reality.

OSH management systems are continually adapting to huge growth in the use of Information and Communication Technology (ICT), as well as rapid changes in the type of technology used, including computer networks, electronic data interchange and the internet. Subsequently, the control of documented information relating to OSH management has increased in importance in supporting the implementation of OSH management arrangements. With OSH information increasingly being kept on devices, such as smart phones and tablets, it is important to allow for management systems information, audit records and other data to be held in this manner and not simply in paper form.

Operation

Step 6 of the MBA program on *Operations Management* covers the development of business processes and the importance of integrating OSH into all aspects of delivery. The operational phase of your OSH management system relates to the implementation of OSH management arrangements. The development of OSH management arrangements is arguably 10% of the work in creating an effective OSH management system, with the other 90% being the execution of associated arrangements. ISO 45001 includes various requirements to ensure that OSH is ingrained within the fabric of your organisation's operations.

In recognition of the broad definition of risk as the effect of uncertainty in the context of achieving organisational OSH objectives, operational controls are required for risks and opportunities that could have a significant impact, be that positive or negative, on the realisation of OSH objectives.

In many organisations, the determination of operational controls will be linked to the risk and opportunity assessment process which will identify, assess and determine control measures for OSH risks and opportunities, in addition to identifying risks and opportunities significant in nature that may require clearly defined operational controls. For example, if your risk and opportunity assessment programme determines that work at height is a significant risk that could have a negative impact on achieving your OSH objectives, then this would inevitably lead to a decision that operational controls would be required.

Operational controls for the management of OSH risks should adopt a hierarchical approach to risk management. This will commence with whether the risk can be eliminated or substituted, to the consideration of engineering controls (such as guarding and physical safeguards), followed by administrative controls (such as permit to work) and finally the use of personal protective equipment. In relation to opportunities, measures will need to be put in place to enhance, exploit or enable the realisation of the opportunity.

 Remember that operational controls will also be needed to realise significant opportunities in addition to mitigating significant risks.

From an operational perspective, it is important that organisations have processes in place for the management of change. Change is a feature of organisational life; however, if it is not managed properly it can result in a deterioration in OSH performance. Changes can include the use of new products, services and equipment, but also changes to work processes, procedures and even organisation structure. An effective change management process should ensure that the responsibilities and authorities for managing changes and their OSH risks and opportunities are identified and appropriately managed.

Another key operational aspect of OSH management system implementation is the fact that the model of a stable workforce, employed directly by an organisation to fulfil a range of functions, has been replaced in many industries by one where staff are only employed to perform core functions. Against this changing backdrop, it is even more important that the OSH risks and opportunities associated with outsourced activities are effectively identified, assessed and controlled. ISO 45001 places considerable emphasis on outsourcing, which many organisations will translate into consideration of controls over contractors and subcontractors, and how far those controls should extend. However, it is worth noting that in some countries there are legally defined demarcations of OSH responsibility between customers and contractors, which would need to be factored into any approach (see Step 6 on *Operations Management* for a more detailed review of the OSH considerations associated with outsourcing, procurement and contractors).

In addition to the focus on day to day operations and associated risks, organisations are required to identify, assess and control emergency risks. Emergency management can touch on numerous related disciplines, such as crisis management and business continuity management. OSH practitioners are increasingly finding that their remit is being extended to encompass these related areas of risk management. A good example of this is security risk management, largely due to the increasing security and political risk profile organisations are facing in some jurisdictions. The growing focus on business resilience is another relevant example (Sheffi, 2015). Although these developments create opportunities for the OSH practitioner, as invariably these approaches require experience in risk based decision making and knowledge of developing and implementing management systems, it is important that suitable training is completed to develop the necessary competencies.

 Avoiding bureaucracy

There is a misconception that an effective OSH management system requires the development of vast amounts of paperwork – policies and procedures. However, it is important to look out for opportunities to streamline OSH documentation and ensure that it is essential and adds value.

A construction company in the Middle East was facing issues with changing behaviours related to work at height, involving the use of a ladder. To address this, they created a one-page document listing the key steps involved in carrying out the activity safely. The script was created in the form of a straightforward work instruction focusing on the safety critical behaviours required at each stage of the task accompanied with photos, diagrams or illustrations. In addition, the document was translated into different languages and incorporated a check list for workers to verify that they have been through the logical steps before commencing an activity.

Basic rules can be created in this fashion for safety critical activities in the workplace, with the key message when scripting key rules to communicate in simple terms the concrete behaviours that are needed to achieve the desired outcome.

Performance evaluation

The implementation of OSH management arrangements needs to be checked to ensure that they are working effectively and generate the desired impact on OSH performance. Within ISO 45001, organisations are required to maintain a process for monitoring, measurement and evaluation, deciding what, when and how these activities need to be performed. Performance evaluation should always be undertaken within the context of OSH objectives determined during the planning stage of your OSH management system.

Performance evaluation should incorporate evaluation of legal and other requirements. Although organisations usually develop comprehensive legal registers which capture legal and other requirements, there is not always an effective process in place to evaluate compliance with these requirements. To strengthen the evaluation of compliance process, it is important to identify the requirement and to specify precisely what is being done in practice to ensure compliance.

The frequency of compliance evaluations will depend on a range of factors, including the importance of requirements, changes in legal requirements and the organisation's past OSH performance.

Periodic internal audits of your OSH management system are important to ensure that it conforms to relevant OSH management requirements and is effectively implemented and maintained. To aid effective internal auditing, ISO 45001 requires an organisation to establish internal audit programmes and define the audit criteria and scope for each audit. It is also essential that those involved in conducting audits are competent to ensure objectivity and impartiality of the audit process.

Whereas an evaluation of legal requirements will establish whether an organisation is compliant with legal requirements, the audit process will determine conformity or fulfilment of requirements within the standard. It is common practice for nonconformities to be categorised as either a major or minor nonconformity, depending on whether the nonconformity relates to the absence of a mandatory process or document, or a weakness within implementation, respectively.

Management review

Top management should review OSH management arrangements at planned intervals to ensure its continuing suitability, adequacy and effectiveness. The need for management review is often a requirement that creates confusion in organisations as it is perceived as regular meetings involving management that may touch on OSH considerations. Importantly, the management review is a comprehensive review of the OSH management system and OSH performance, with clearly defined inputs and outputs to enable decisions to be made regarding the future direction of OSH management.

Notwithstanding, the OSH management review is often performed as part of a wider management review meeting that may address a range of different topics, such as corporate strategy, financial and commercial performance, business function activity and performance and stakeholder management.

Improvement

The final part of ISO 45001 deals with improvement. Improvements to your OSH management system and OSH performance can be identified reactively and proactively. The reactive dimension relates to the implementation of corrective actions following an incident or non-conformity. The active, or pro-active, dimension applies when observations or opportunities for improvement are identified that may not have led to an incident or non-conformity, enabling the identification of potential improvement actions.

It is important that all organisations have robust and comprehensive arrangements in place for the management of incidents and non-conformities, including reporting, investigating and taking action. Furthermore, a process to foster continual improvement is needed to ensure that organisations continually improve the suitability, adequacy and effectiveness of the OSH management system. A focus on OSH innovation can help to ensure that sufficient time and energy is devoted to continual improvement (see Step 9 on *Innovation and Entrepreneurship* for discussion on different approaches to foster a culture of continual improvement).

Summary

Transitioning into a general management role typically requires a broader perspective, as leading a business and managing day-to-day operations needs to be informed by a strategic view across all core business activities and support functions. In this sense, to be an effective General Manager you will need to possess leadership and management skills, such as inspiring and motivating people, recognising when change is needed and executing decisions in the best interests of the wider organisation.

If you do aspire to become a General Manager in your organisation, or another organisation, then it is important that you understand management systems, incorporating the various processes and tools that define an organisation's ways of working. The introduction of ISO 45001 is a significant landmark for the OSH profession and will bring about changes in OSH management systems and ensure that OSH is more deeply embedded at a strategic, operational and functional level. The training and experience that many OSH practitioners receive in management systems provides a good grounding in general management principles and if this can be coupled with an understanding in strategy, marketing, finance, people management and organisational change, then you are certainly heading in the right direction to become a credible General Manager.

Key takeaways

- General Management is concerned with the leadership and management of an organisation as a whole – *are you getting the opportunities, experiences and training to develop your general management capabilities?*
- There are differences between leadership and management, with leaders providing vision, inspiration and change and managers seeking order, control and rapid resolution of problems – *do you recognise the distinctions and commonalities between OSH leadership and management?*
- General Management within the context of OSH relates to the development, implementation and maintenance of an OSH policy and management arrangements – *is your OSH management system achieving the desired results?*
- Planning is fundamental to successful OSH management; however, there should be a shift to looking at opportunities in addition to risks – *have*

you identified the opportunities that could have a positive impact on OSH objectives?

- The publication of ISO 45001 will introduce new language and concepts to the discipline and profession of OSH – *is your organisation ready for the transition towards ISO 45001, particularly the increased focus on stakeholder management and leadership?*

References

Abu Dhabi Occupational Safety and Health Centre (OSHAD). (2016). *Safety in heat.* Retrieved from www.oshad.ae/safetyinheat/en/index.php.

Berry, E. (2015). *Pestle analysis essentials.* Seattle, WA: CreateSpace, Independent Publishing Platform.

Health & Safety Executive (HSE). (2015). *Leading and managing for health and safety.* Retrieved from www.hse.gov.uk/managing/leading.htm

Health & Safety Executive (HSE). (n.d.). *Why leadership is important.* Retrieved from www.hse.gov.uk/leadership/whyleadership.htm.

International Organization for Standardization (ISO). (2018). *ISO 45001: 2017 Occupational health and safety management systems – requirements with guidance for use.* Geneva: ISO.

Sheffi, Y. (2015). *The power of resilience: How the best companies manage the unexpected.* Cambridge, MA: MIT Press.

Stadler, S. (2015). How to become a CEO: These are the steps you should take. *Forbes.* Retrieved from www.forbes.com/sites/christianstadler/2015/03/12/how-to-become-a-ceo-these-are-the-steps-you-should-take/#40f5211267cd.

Ward, S., & Chapman, C. (2011) *How to manage project opportunity and risk: Why uncertainty management can be a much better approach than risk management.* Hoboken, NJ: Wiley, 3rd edition.

3 Corporate social responsibility and governance

You have reached Step 3 of the programme, the final Step linked to the first key theme of the book on the need to "be strategic". In this Step, the concepts of Corporate Social Responsibility and Governance will be discussed, along with related areas including Social Accountability, Sustainability and Enterprise Risk Management. The purpose of this Step is to provide the OSH practitioner a greater understanding of emerging corporate strategy and management topics. These considerations relate to the roles of the Chief Executive Officer (CEO) or Managing Director and the Board of Directors and how the OSH practitioner can assist in steering the organisation in the right direction on important decisions.

On completion of Step 3, you will be able to:

- Understand what is meant by corporate social responsibility and its association with OSH
- Recognise the importance of ethical decision making in business and on an individual basis
- Understand the relevance of sustainability and social accountability to OSH
- Explain the importance of integrating OSH within a wider enterprise risk management framework
- Understand corporate governance and the components of an effective corporate OSH governance framework.

Corporate social responsibility and ethics

All organisations want leaders who say and do the right thing. However, the short term financial pressures faced by many companies often result in decisions where profits are put above OSH, environmental and other non-financial considerations. In a world that is centred on money and the typical life-span of a CEO is often less than one year, it is perhaps not surprising that organisations become preoccupied with financial decisions; however, your role as an OSH practitioner is to ensure that there is an appropriate balance between financial pressures and OSH considerations.

Although the primary purpose of any organisation is to provide a social-economic contribution, it is important to remember that whilst shareholders

74% of respondents either "agreed", or "strongly agreed" that Corporate Social Responsibility is an important topic in my role as an OSH practitioner.

expect value maximisation through profitability, a sustained operation with economic prosperity of growth is typically the primary objective of shareholders. In this sense, a commitment to OSH management is fundamental to the creation of a sustainable business model.

The terms sustainability and Corporate Social Responsibility, or CSR, are often used interchangeably and government and stakeholder interest in this area in many jurisdictions continues to increase. Fundamentally, CSR is about an organisation considering and taking account of the present and future impacts of its decisions on society and the environment. Invariably, this will involve giving something back to the society, economy and environment through voluntary initiatives. This commitment and focus can help to drive ethical behaviours (SHP Online, 2014).

Retrospectively, decisions made in cases such as the Volkswagen incident may appear baffling, but they indicate the pressures that many organisations are under

 ### *Ethical decision making*

Short term financial pressures can often result in situations where organisations do not invest enough time and money in OSH management and can be inclined to make decisions which are not always in the wider interest of stakeholders.

Take the case of Volkswagen in September 2015, when the German engineering behemoth admitted to US regulators that it programmed its cars to detect when they were being tested and altered the running of their diesel engines to conceal their true emissions.

Let's imagine that in your role as OSH Manager you became aware of these violations. You have an opportunity to influence key decision makers to do the right thing – what would be your approach?

Ask yourself, what would you do to try and influence a different response? Also, what changes would you make to improve future decision making on ethical matters?

to deliver financial returns and the impact this has on trying to rationalise certain decisions. Another famous case on ethical decision making that has relevance to OSH management is that of the Ford Pinto (Gioia, 1992). This model was prone to explosion in the event of a rear-end collision due to a design fault with the petrol tanker being positioned at the rear of the vehicle. A memo leaked from the factory that Ford had undertaken a cost-benefit analysis, comparing the cost of conducting a vehicle recall against the likely number of incidents that might occur and the associated compensation payouts, before deciding not to carry out a recall. Again, on reflection, this is a case which appears incredulous but highlights the financial pressures that many organisations face and willingness to put aside ethical considerations. As indicated at the start of this Step, although as an OSH practitioner you may not reach an Executive level position, it is important that you play your role in steering these Executives and the wider organisation in the right direction.

As an OSH practitioner, it is also important to be aware of how ethical conduct relates to your own behaviours. Some OSH professional bodies, such as IOSH, have a code of conduct that places expectations on the actions on OSH practitioners (IOSH, 2013), and although at times there may be temptations to engage in unprofessional behaviour, ethical behaviours are paramount in maintaining the credibility of our profession and the trust of the people that we work with.

Sustainability and social accountability

Many companies are actively integrating sustainability principles into their business and there is evidence that these organisations are pursuing goals that go far beyond reputational considerations, with sustainability becoming a key factor within strategic planning, supply chain management and budgeting (Bonini, 2011). It is clear that sustainability and business now go hand-in-hand, with sustainability regarded as a far-reaching subject, addressing economic, social and financial considerations that relate to the ability of an organisation to deliver value.

Many organisations around the world report on sustainability performance in line with standards developed by the Global Reporting Initiative (GRI). GRI is an international independent organisation that helps organisations understand and communicate the impact of business on critical sustainability issues, such as climate change, human rights, corruption and many others, with 92% of the world's largest 250 corporations reporting on their sustainability performance (GRI, n.d.). The production of sustainability reports provides organisations an opportunity to publish information about the economic, environmental and social impacts caused by their activities. These reports also help to present an organisation's values and governance model and demonstrate the link between corporate strategy and the commitment to a sustainable global economy.

The links between sustainability and OSH are well documented, with the American Society of Safety Engineers (ASSE) defining OSH Sustainability as 'The responsibility to ensure that the protection of human life and the safety, health and well-being of workers, customers, and neighbouring communities are among the primary considerations in any business practices, operations or development' (ASSE, 2010). This definition is focused on health, safety and welfare of

Table 3.1 Elements of the Occupational Safety and Health Sustainability Index

Aspects	Elements
Values and beliefs	• OSH social responsibility commitment
	• Codes of business conduct
Operational excellence	• Integrated and effective OSH management systems
	• Professional OSH competencies
Oversight and	• Senior leadership oversight of OSH
transparency	• Transparent reporting of key OSH performance indicators

all stakeholders, with six elements detailed in the ASSE OSH Sustainability Index as shown in Table 3.1:

The social component of sustainability is where significant cross-over points exist with OSH. Indeed, a useful standard that OSH practitioners should be aware of is the Social Accountability, SA 8000:2014 standard (Social Accountability International, 2014). This is one of the world's first auditable social certification standards for workplaces, across all industrial sectors. The standard is based on the UN Declaration of Human Rights, conventions of the ILO, UN and national laws, spanning industry and corporate codes to create a common language to measure social performance.

Enterprise risk management

In recent years, there has been a continuing shift to risk management, as opposed to safety management. This shift has been driven by financial crises, which have largely been caused by weak governance and a lack of financial regulation, all of which have brought the subject of risk management firmly onto the agenda of most large organisations.

Enterprise Risk Management (ERM) is the methods and processes used by organisations to manage risks and opportunities that may have an impact on the realisation of business objectives (Hampton, 2009). Re-defining risk in ISO 45001 is a significant first step in integrating OSH management within wider business frameworks for risk management. After all, the approach to managing OSH risks should be no different from any other risk in an organisation, in that the risk should be identified, assessed, controlled and monitored.

This systematic approach to the management of risks is reflected in the philosophy of ISO 45001 and other guidance that apply to risk management. For example, the Financial Reporting Council Guidance (formerly the Turnbull Report) recommends that listed companies have robust systems of internal control, covering not just financial risks but also risks relating to the environment, business reputation and OSH. In this sense, ERM provides a far broader lens for the management of risk (Financial Reporting Council, 2014).

In many respects, OSH practitioners are well positioned for the continuing transition to risk management. Many of the skills used by OSH practitioners in

Enterprise risks

An effective ERM framework will encompass the management of risks and opportunities at a strategic, functional and operational level. Invariably, this will involve inputs from various departments and people at all levels of the organisation.

Spend some time thinking about the different parts of your organisation using the guide provided to help.

Figure 3.1 Examples of enterprise risks

What are the key risks and opportunities? Are there any significant inter-actions and interdependencies that may arise in relation to OSH risks and opportunities?

the identification and assessment of risk will be needed when considering wider risks faced by an organisation. However, in many cases, greater work is needed by OSH practitioners to understand financial risks, a capability which could potentially help OSH practitioners in reaching the status of Chief Risk Officer.

Risk management strategies

As previously indicated, the role of the OSH practitioner has developed to incorporate loss prevention though a range of risk management strategies. Risk control strategies may be classed into four main areas (IOSH, 2015):

Risk avoidance – avoiding a risk by discontinuing the operation producing that risk.

Risk retention – managing the risk within the organisation, with any loss arising from poor risk management being totally financed from within. This option may be followed consciously or unconsciously – it's what happens if risks aren't fully identified.

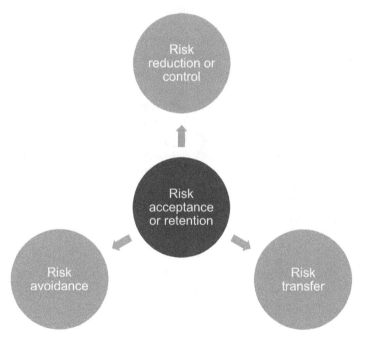

Figure 3.2 Risk management strategies

> Risk transfer – involving the assignment of the costs of certain potential losses from one party to another. The most common way of doing this is by insurance, but other forms of contractual risk transfer include sales contracts and employing third parties, e.g., contractors.
>
> Risk reduction – managing the risk in the organisation by implementing a programme designed to protect the organisation's assets from wastages caused by accidental loss.
>
> Try to ensure that your approach to OSH risk management is aligned with broader business risk management strategies that may be implemented in your organisation.

Governance

As discussed in Step 2 on *General Management*, the OSH practitioner's role invariably focuses on the development, implementation and maintenance of OSH management systems and processes. It is the execution of these management processes that plays a critical role in preventing workplace incidents and ill health. However, although the OSH practitioner may contribute significantly to OSH management-related decisions, they are not always given the opportunity to influence governance processes, which sit above OSH management provisions.

Governance is the system by which organisations are directed and controlled by their Board of Directors. It is through the lens of governance that the broadest strategic decisions are taken and management is held to account. Governance processes are somewhat distinct from management, which are often regarded as the day-to-day decisions taken to run a business.

The term "corporate governance" has become common parlance (Cadbury Report, 1992), founded on the belief that risks should be systematically identified, assessed, controlled and reviewed (Turnbull Report, 1999). Indeed, with the growing recognition that corporate governance arrangements need to encompass all business risks, OSH is now increasingly being seen by organisations as a corporate governance issue.

 Remember, governance relates to high-level processes for direction and control, whereas, management relates to the day-to-day decisions in running a business.

Corporate OSH governance

Guidance produced by the HSE provides an understanding of how OSH fits within traditional corporate governance and the benefits an integrated approach can bring (Health & Safety Executive, 2006). Importantly, the guidance details what corporate governance as applied to OSH may look like in practice, presenting a conceptual framework made up of seven key principles, which are discussed in the following sections.

Director competence

A key pillar of an effective framework for corporate OSH governance is director competence. A good starting point to encourage top management to take an interest in OSH is to recommend that they attend training. There are various internationally-recognised OSH training programmes that have been developed to address the specific requirements of top management. For example, IOSH Leading Safely is ideal for top management in small and large organisations respectively. This type of programme can help to raise awareness of the need for top management to provide sound strategic direction for OSH and the importance of their role in shaping the OSH culture in the organisation.

In addition to training, the OSH practitioner can make top management aware of industry conferences and external seminars that address OSH, or recommend involvement in government committees. These forums provide an opportunity for top management to engage in OSH debates and may help in enhancing interest in the discipline. To further develop levels of awareness, it is also commonplace in some organisations for external speakers to deliver brief presentations on topical OSH issues at board meetings, or to permit an external OSH advisor to sit on board sub-committees dedicated to risk management.

 Make sure to select a high quality training provider for any OSH training provided to senior leadership, as if the training goes badly it can seriously undermine the implementation of your OSH strategy.

Director roles and responsibilities

Unfortunately, a belief can exist in some organisations that OSH is regarded solely as the remit of an OSH department and is not the responsibility of top management. However, integrating OSH into holistic corporate governance structures is a constructive way of promoting collective responsibility from top management.

Moreover, guidance published by the HSE, aimed specifically at directors, offers specific advice on their OSH responsibilities (Health & Safety Executive, 2003a). One specific action point in the guidance recommends that the board of directors needs to formally and publicly accept its collective role in providing OSH leadership in its organisation. Although this guidance was superseded (Health & Safety Executive, 2013), the message remains the same in that, irrespective of the appointment of an OSH director, members of the board should have both individual and collective responsibilities for OSH.

The problem with responsibilities is that they tend to be delegated. Subsequently, it is important that the OSH practitioner, in liaison with human resources professionals, ensures that job descriptions for all positions in an organisation, including top management, include responsibilities and accountabilities. The preparation of position descriptions with applicable jurisdictional accountabilities can be an effective way of ensuring that top management fully understand their commitment to providing effective OSH leadership.

Culture, standards and values

Fundamentally, OSH is about people. It is about preventing death, injury and ill health to people at work and those affected by work activities. However, this laudable goal will never be achieved if there is a misalignment between OSH requirements and the core values of the organisation. Top management commonly proclaim organisational visions and values (the concept of a business vision has been explored in Step 1 on *Strategy and Leadership*), with values relating to the core principles or standards in a business – summing up what the business stands for and what makes it special. For example, the business values at the Coca-Cola Company include leadership, passion, integrity, accountability, collaboration, innovation and quality (Coca Cola, n.d.). It is important for the OSH practitioner to be aware of the values of the organisation and determine their compatibility with the overall vision for OSH.

 Take time to identify and understand the values within your organisation and ensure OSH messages are aligned.

The tone that is set from the top of the organisation is vital in driving OSH improvements, with evidence suggesting that management styles characterised by openness and flexibility are the most effective in promoting a positive OSH culture (Health & Safety Executive, 2003b). To encourage greater commitment from top management, the OSH practitioner can play a role as facilitator in providing opportunities for top management participation in OSH. This can include organising periodic management OSH tours or encouraging top management to be involved in OSH competitions. These initiatives can also help engender commitment from lower levels of the organisation and assist in cascading OSH messages throughout the organisation.

Strategic implications

Governance incorporates various strategic implications, including setting direction, making policy and strategy decisions, overseeing and monitoring organisational performance, and ensuring overall accountability. Indeed, a comprehensive

OSH corporate governance framework should run hand in hand with your OSH strategy. Step 1 on *Strategy and Leadership* provides extensive discussion on defining and executing OSH strategy and the importance of establishing the OSH future you desire in terms that employees can understand.

Within an effective OSH governance model, the board of directors should be responsible for driving the OHS agenda, understanding the risks and opportunities associated with OSH matters and any market pressures which might compromise the values and standards, and ultimately establishing a strategy to respond. Invariably, decision making of this nature will involve input from OSH practitioners and other specialist advisors.

Performance management

Once an OSH vision has been established, it is important for the board to be involved in setting objectives and targets to meet the overall vision. The OSH practitioner should ensure that the organisation's approach to measuring OSH performance encompasses lagging indicators or measures of failure, for example, number of lost time accidents (LTAs), days lost due to occupational ill health, etc., and leading indicators that provide a more holistic perspective of the OSH culture and effectiveness of OSH management arrangements, for example, the percentage of top management attending OSH training, percentage of corrective action requests closed out, etc.

Top management often refer to the 'balanced scorecard' when discussing performance management (Kaplan & Norton, 1992). This tool essentially requires a combination of financial (revenue growth, profitability) and non-financial (customer service, sustainability) metrics to be utilised when measuring an organisation's performance. Although it is argued that OSH has financial benefits, in the context of the balanced scorecard it would typically be regarded as a non-financial consideration. The OSH practitioner should recommend that OSH metrics are given sufficient weighting within methodologies of this nature.

Internal controls

The board should ensure that existing internal control structures provide for the identification and management of all key risks, including OSH. The OSH practitioner may be regularly involved in performing audits of OSH management arrangements; however, it is important to ensure that the approach to OSH auditing is consistent with other internal audit processes within the business. Furthermore, even if an organisation does not follow ISO 45001 requirements, it is good practice to ensure that top management contributes to a periodic review of OSH performance.

One of the issues with internal control mechanisms and associated board (group) decision making is 'group think' (Janis, 1972), the phenomenon whereby members of the group are so concerned in striving for consensus that alternative or conflicting perspectives are often ignored, irrespective of their merits. The role of the OSH practitioner should include the power to be able to offer opinion on important decisions with OSH implications prior to approval from top management. Indeed, the word governance is derived from the Greek word *kubernetes*, which means 'helmsman of the ship' (Garratt, 2005), and even though the OSH practitioner may not decide the direction that the ship takes, you can certainly help in manning the rudder.

Organisational structures

The final element of an effective corporate OSH governance framework is organisational structures. Previous discussion on director roles and responsibilities emphasises that although directors may have individual responsibilities for OSH they should operate collectively and that OSH is an issue for the whole board. Moreover, integrating OSH into holistic corporate governance structures is a constructive way of promoting collective responsibility from top management. From the OSH practitioners' perspective, it is important that OSH requirements are communicated to top management and that careful attention is paid to ensuring co-ordination between the systems and processes developed for managing risks throughout the business.

There remains a growing trend for OSH to be part of a wider enterprise, risk-management perspective, placing significant emphasis on effective organisational design. This shift has resulted in organisations commonly establishing separate board sub-committees to consider risks which may affect the business, at a corporate, functional and project level. Although the creation of an executive OSH committee may not be warranted, it is important that the board is organised to deliver OSH governance and that appropriate structures are put in place to make it happen.

The issue of organisational design also raises questions as to where the OSH function should sit in an organisation and associated reporting lines. In some organisations, the OSH function reports into an Operations department, and others into a Human Resources department. For OSH to be positioned as an impartial voice, it is ideal for the OSH function to report directly to General Management, or the Managing Director. In this sense, there is less risk of OSH decisions being impacted by conflicts of interest that may arise under alternative reporting arrangements.

Summary

Although governance processes may vary across different jurisdictions and business ownership structures, it is important that the OSH practitioner understands the importance of corporate occupational OSH governance and some of the ways to encourage top management involvement in OSH. Furthermore, with the increasing need for OSH practitioners to have the ability to talk to business units in strategic terms, attempts should be made to propose solutions that are integrated with wider business considerations. Demonstrating an understanding

Creating an OSH governance framework

Research published by the UK HSE has identified seven principles of best practice relating to corporate governance. The relationship between these elements is depicted in the figure shown and discussed in this Step of the programme.

How would you apply this in your organisation?

What are the gaps in your current approach to corporate OSH governance?

Figure 3.3 Corporate Occupational Safety and Health governance framework

of wider governance issues and recognition of the business realities of decision-making can help improve top management's perception of the role of the OSH practitioner and help ensure that OSH requirements are given the attention they deserve. OSH practitioners should look to avoid fear-mongering and attempt to motivate top management to take a leading role on these matters.

Research shows the benefits of governance processes as applied to OSH, including long-term prosperity and value creation (Health & Safety Executive, 2006). However, more needs to be done to effectively integrate OSH into the existing governance framework and wider arrangements for OSH management in many organisations, which represents a timely segue into the next key theme within the 10 Step MBA program of the need for OSH practitioners to "be cross-functional".

Key takeaways

- Corporate social responsibility is about giving something back to the economy, society and environment – *do you have a strategic approach to CSR, incorporating voluntary initiatives?*
- Ethics is about doing the right thing, which can be challenging is an environment driven by financial pressures – *do you have the balance right between OSH and finance?*
- The links between sustainability and OSH are often blurred; however, OSH is a key consideration of the social element of sustainability – *do your sustainability strategies and programs sufficiently address the social component?*
- Enterprise risk management adopts a holistic approach to risk management at a strategic, functional and operational level – *is there an effective level of integration between OSH and ERM?*
- There is a strong relationship between corporate governance and OSH, centred on the principles of sound risk management – *do you have an effective approach to corporate OSH governance?*

References

American Society of Safety Engineers (ASSE). (2010). *Safety and health sustainability index taskforce*. Working paper. Park Ridge, IL: American Society of Safety Engineers.

Bonini, S. (2011). *The business of sustainability: McKinsey global survey results*. McKinsey&Company. Retrieved from www.mckinsey.com/business-functions/sustain ability-and-resource-productivity/our-insights/the-business-of-sustainability-mckinsey-global-survey-results.

Coca Cola Company (CCC). (n.d.). *Mission, vision and values*. Retrieved from www.coca-colacompany.com/our-company/mission-vision-values.

Financial Reporting Council (FRC). (2014). *Guidance on risk management, internal control and related financial and business reporting*. Retrieved from www.frc.org.uk/Our-Work/Publications/Corporate-Governance/Guidance-on-Risk-Management-Internal-Control-and.pdf.

Garratt, B. (2005). Can boards of directors think strategically? Some issues in developing direction-givers' thinking to a mega level. *Performance Improvement Quarterly*, 18(3), 26–36.

Gioia, D. A. (1992). Pinto fires and personal ethics: A script analysis of missed opportunities. *Journal of Business Ethics*, 11(5), 379–389.

Global Reporting Initiative (GRI). (n.d.). *About GRI*. Retrieved from www.globalreport ing.org/information/about-gri/Pages/default.aspx.

Hampton, J. (2009). *Fundamentals of enterprise risk management: How top companies assess risk, manage exposure, and seize opportunity*. Amacom.

Health & Safety Executive (HSE). (2003a). *Director's responsibilities for health and safety*. Sheffield, UK: HSE Books, INDG 343.

Health & Safety Executive (HSE). (2003b). *The role of managerial leadership in determining workplace safety outcomes*. Sheffield, UK: HSE Books, Research Report 044.

Health & Safety Executive (HSE). (2006). *Defining best practice in corporate occupational health and safety governance*. Sheffield, UK: HSE Books, Research Report 506.

Health & Safety Executive (HSE). (2013). *Leading health and safety at work.* Sheffield, UK: HSE Books, INDG 417, C700.

Institution of Occupational Safety and Health. (IOSH). (2013). *Business risk management: Getting health and safety firmly on the agenda.* Leicester, UK: IOSH, POL3728.

Institution of Occupational Safety and Health (IOSH). (2015). *Code of conduct, guidance and disciplinary procedure.* Leicester, UK: IOSH, COR1081.

Janis, I. L. (1972). *Victims of groupthink: A psychological study of foreign-policy decisions and fiascos.* Boston, MA: Houghton Mifflin Company.

Kaplan, R., & Norton, D. (1992). The balanced scorecard – measures that drive performance. *Harvard Business Review,* 70(1), 71–79.

Report of the committee on the financial aspects of corporate governance: Cadbury report. (1992). Gee. A division of Professional Publishing Ltd.

SHP Online. (2014). *Health and safety and the CSR/sustainability agenda.* Retrieved from www.shponline.co.uk/health-and-safety-and-the-csr-sustainability-agenda/.

Social Accountability International (SAI). (2014). *Social Accountability 8000 International Standard.* New York, NY.

Turnbull, N. (1999). *Internal Control: Guidance for Directors on the Combined Code.* London, UK: Institute of Chartered Accountants in England and Wales.

Part II

Cross-functional

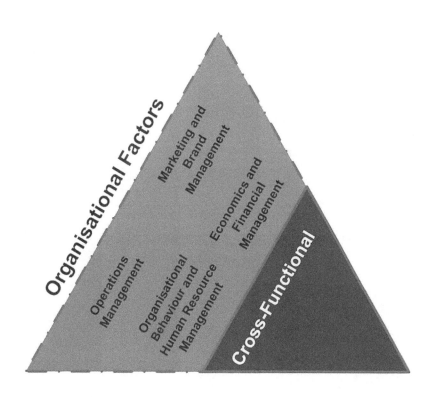

4 Organisation behaviour and human resource management

Welcome to Step 4! The goal of achieving excellence in OSH performance is never easy as it involves dealing with people! All organisations are essentially created and designed by people to fulfil human objectives. Subsequently, an understanding of human behaviour in organisations is essential in achieving positive OSH outcomes. The management of human behaviour in organisations requires a collaborative approach with input from various functions, which links to the second of our three themes in the OSH Practitioner Transformation Model on the need to "be cross-functional". The interaction between Human Resources (HR) and OSH is arguably the most critical of these relationships with other organisational functions and will be considered in this Step with a focus on the factors required to develop an effective relationship.

On completion of Step 4, you will be able to:

- Understand the concepts of organisational culture and OSH culture
- Appreciate the importance of cultural maturity when selecting OSH interventions
- Recognise the relevance of emotional, rational and situational factors to the design and implementation of behaviour change programs
- Identify the need for worker consultation and involvement in developing a positive OSH culture
- Determine the opportunities for collaboration that exist between HR and OSH functions.

A people-centric approach

This Step of the program is particularly important because the role of many senior OSH practitioners has substantially changed from a technical specialist to a position that involves working closely with people at all levels of an organisation. In this respect, the support offered by the HR function and the creation of a strategic alliance between HR and OSH functions is essential to the development of positive OSH standards, culture and performance.

83% of respondents either "agreed", or "strongly agreed" that to do my job well, I need to understand the human resources function in my organisation.

HR and OSH are fundamentally people-centric functions concerned with shaping organisational culture to achieve business success. Before discussing the cross-over points and collaborative opportunities that exist between HR and OSH, it is necessary to provide some introductory discussion on organisational culture and the increasing focus on changing attitudes and behaviours that is evident in many organisations.

Organisational culture and OSH

Organisational culture is a system of shared values and beliefs about what is important within an organisation, what behaviours are appropriate and about feelings and relationships both internally and externally. It is recognised that an organisational culture can be created to promote OSH, commonly referred to as an OSH culture (Roughton & Mercurio, 2002).

Cultural maturity

The OSH culture present in an organisation may have differing levels of maturity. The concept of cultural maturity in the field of OSH is related to the safety culture maturity model (Fleming, 2001). The safety culture maturity model suggests that organisational culture, with respect to OSH, progresses sequentially through five stages. At level one of the model, the safety culture is expressed as *emerging*, with OSH being defined in terms of technical and procedural solutions and compliance with legislation. The OSH department is perceived to have responsibility

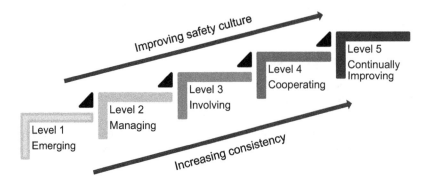

Figure 4.1 Safety culture maturity model (registered trademark of the Keil Centre Limited)

for OSH and accidents are unavoidable and part of the job. Most frontline staff are uninterested in OSH and may only use it for the basis of other arguments, for example shift changes. The safety culture maturity model progresses through several stages from managing, involving and co-operating until it reaches the final stage of continual improvement.

At the final stage of maturity (continual improvement), the prevention of all injuries or harm to employees, both at work and at home, is a core company value. The organisation is constantly striving to find better ways of improving hazard control mechanisms. All employees share the belief that OSH is a critical aspect of their job and accept that prevention of non-work injuries is important and are empowered to address OSH problems in the workplace and take ownership of these issues. The question for many OSH practitioners is, how can I bring about this step-change in my organisation?

Well, guidance states that management commitment and style, training and competence, communication, compliance with procedures, organisational learning and employee involvement are important factors in developing a positive OSH culture (Health & Safety Executive, 2002). However, the growing recognition that improving OSH performance requires a focus on shaping of workplace behaviours has led to a proliferation of Behaviour Based Safety (BBS) programs, particularly in the major hazard industries. However, it is important that your organisation has the right level of maturity for a behaviour based approach. Importantly, you need to get the basics right, in terms of technical and administrative controls, before you spend too much time looking at behaviour.

Behaviour

To improve OSH performance and reduce occupational injury and ill health, a focus is invariably required on developing and reinforcing positive protective behaviours. However, influencing human behaviour is complicated. A commonly adopted idiom indicates that there are two basic methods that can be employed to shape behaviour – "Carrot and Stick" (Dickenson, 2001). The Carrot is intended to reward individuals who behave in accordance with recognised norms and standards, through various extrinsic (monetary) and intrinsic (personal – i.e. peer recognition) means. The stick is designed to punish and typically employs the use of reprimands, such as disciplinary action to achieve desired behaviours. Unfortunately, human behaviour is not quite as simple and two-dimensional as suggested by the Carrot and Stick model.

A multi-pronged approach

Recent work on organisational and individual behaviour and change management suggests that a multi-level approach needs to be embraced which considers emotional, logical and situational considerations. The combination of these components can help to formulate an excellent recipe for change management (Heath & Heath, 2010). It is worthwhile spending some time discussing these considerations and their relevance in the design and implementation of workplace OSH behavioural change strategies.

 There is no quick fix to changing behaviour. Ensure that your OSH behaviour change programs address a range of emotional, rational and situational factors.

Emotional factors

The emotional part of our brain is widely regarded as more powerful than our rational side with OSH change management initiatives commonly referring to the need to reach the "Hearts and Minds" of the workforce (Katzenbach, 2000). This dichotomy illustrates the brain having two systems: system one controls our intuitive responses; and system two relates to our calculated and more objective considerations of a stimulus or situation (Kahneman, 2011). This understanding heightens awareness of the need to try to connect with people on a personal level and to ensure that behaviour change efforts are perceived as achievable.

Consider the challenge that many organisations face in implementing OSH management arrangements, including compliance with site rules and procedures. Although worksite OSH protocols are important they rarely appeal to our emotional side. Consequently, this can create a disconnect between the procedures within the management system and the desire to enact the necessary behaviours to ensure compliance. Furthermore, time and money can be invested in training on management arrangements, but unless interventions relate to workers on a personal level these individuals may continue to exhibit unsafe behaviours. To address these types of challenges, sometimes a subtler approach is required, to address our emotional inclinations.

Raising awareness of the need to follow OSH procedures amongst migrant workers, who may have low levels of literacy and OSH understanding, is a relevant example that OSH practitioners increasingly face. Behavioural change in this context could be supported using industrial theatre, or other media, focusing on a family who are struggling to cope emotionally and financially after a parent is killed in a workplace incident. Many migrant workers send remittances home to family members every month and perhaps if they can visualise the impact that a workplace incident could have on their loved ones then this could help to encourage adoption of the necessary safe behaviours. If messages of this nature are reinforced on a regular basis it can help workers begin to sense intuitively the importance of OSH.

Creating realistic OSH goals

We can quickly become demotivated if OSH goals are perceived as unrealistic. Nevertheless, it is important to set ambitious yet achievable goals when striving for excellence in OSH performance. With any behavioural change effort, it is important that the change is broken down into manageable phases to instil the confidence among the workforce that the goal is realistically attainable. Furthermore, to help appeal to our emotional side, it is critical to emphasise how far you have come, as opposed to how far you have to go. This will help to create a sense of accomplishment and generate the impetus that is needed to improve further (see Step 1 on *Strategy and Leadership* for further discussion on the creation and pursuit of OSH goals).

Thinking of yourself and others

It is important that behaviour based programs facilitate workers' understanding that they have a responsibility to keep safe and healthy for their families. This can be extended further in relation to the need to recognise that we are often part of a larger workforce which represents one big family. It is important to develop the mentality that we would not want anyone within their fraternity or sorority to be injured at work. The observation of safety critical behaviours and subsequent feedback in terms of safe and unsafe behaviours can help people appreciate that they do not just have a responsibility to themselves and their dependents, but also to their extended family (i.e. colleagues at work). This can be supported by team-based incentives, or collaboration awards, whereby to receive incentives at periodic intervals, no injury would be acceptable to anyone within nominated teams.

Rational factors

Up until now we have been looking at the emotional part of decision-making; however, self-control typically involves the rationalization of an emotive reaction, evaluating alternatives and adopting a reasoned response. The rational side of the part of the brain sets the alarm clock 30 minutes early at night, to wake up in time to avoid the morning traffic. However, it is the emotional part of the brain that wakes up in the comfort of a warm bed and presses the snooze button, resulting in the eventual need to drive at excessive speed to arrive at work on time! Clearly, behavioural change programs also need to allocate sufficient time to appeal to our rational side.

Focus on the positives

As human beings, we have a psychological tendency to focus on the negative as opposed to the positive. If your organisation has achieved excellent OSH performance on projects throughout a region, but has one project with unsatisfactory performance, greater attention will invariably be given to the project that is struggling. We should spend more time focusing on what is working as opposed to what is not. For example, in relation to project OSH performance, questions to raise could include: "what are we doing well on other projects?", "Why is this the case?" and "Can these approaches be transferred to underperforming projects?".

Many organisations employ benchmarking techniques to compare OSH performance, both internally and externally. However, try to ensure that sufficient focus is given to successes that are bred within the organisation. The development of a knowledge management system is one way for organisations to identify and institutionalise bright spots, whereby lessons from company projects and other activities can be shared throughout the organisation. These types of initiatives should not only document problems encountered, but represent the opportunity to communicate successes that could be replicated in other parts of the organisation.

 Try to avoid scare-mongering with OSH behaviour change programs; focus on the positives and sell the benefits of effective OSH management.

Script the critical moves

OSH management systems incorporate procedures detailing necessary behaviours for undertaking work activities, particularly safety critical tasks. However, as previously indicated, a divide can occur between the management system and the implementation of desired behaviours. This problem can be even more apparent in working environments where there are many migrant workers with varying educational qualifications and work experience, due to challenges in communicating and understanding documents, in consideration of different languages and cultures (Grivna et al., 2012). To address these problems, it is important to provide workers with clear behavioural direction in the form of "Simple Rules" (Eisenhardt & Sull, 2001).

Point to the destination

To achieve positive OSH behaviours, the workforce needs to understand the goals in relation to OSH performance. This direction provides the foundation for fostering appropriate behaviours and developing a positive OSH culture. As discussed in Step 1 on *Strategy and Leadership*, the creation of a positive and concrete vision for OSH is critical in providing clarity. To appeal to our rational side, any vision should provide clearly defined criteria for achieving this goal.

Situational factors

The Health & Safety Executive (1999) propose that the environment or organisation in which we operate is one of the most significant influences on safe behaviour. Peer pressure to subscribe to certain unsafe behaviours (e.g. not wearing items of personal protective equipment) represents a common example of how our environment can shape our behaviour. There are a range of techniques that can be employed to address the situational factors that may influence behaviour.

Change the environment

It is important to recognise that changes to the environment can often eliminate or mitigate the likelihood of unsafe behaviours. A relevant example would be work at height activities. Work at height, due to its inherent risks, should be avoided where possible. However, if work at height cannot be avoided, it is far better to focus on the environment through the introduction of engineering controls that prevent individuals falling from height than to rely solely on behavioural interventions, such as training and supervision.

Fundamental attribution error

Incentives typically assume that undesired behaviours are attributable to the individual, a phenomenon known as the Fundamental Attribution Error (Gilovich & Eibach, 2001).

Consider an activity that involves feeding material into a piece of machinery. The equipment is fitted with interlocking guards that are required to be in place before the process can commence. Nevertheless, workplace inspections have identified that workers are regularly bypassing the interlocking device to increase productivity.

What would you do in this situation? A common approach would be to introduce some form of incentive (Carrot or Stick) to change behaviour. However, what if further investigation determines that the root cause of the unsafe behaviour is time pressure from management, resulting in a tendency to bypass safety devices to speed up the process?

This is a classic example of the Fundamental Attribution Error in practice, in that it is the environment and not the individual that is responsible for the unsafe behaviour.

Consider the challenge that regulators and organisations are facing around the word in preventing mobile phone use when driving. Various behavioural change campaigns have been introduced with limited impact as unfortunately people still have the capability, opportunity and motivation to exhibit unsafe behaviour (Michie et al., 2011). In this situation, one way to approach behaviour change is to tweak the environment through the introduction of technology that makes it physically impossible to use your mobile phone when driving.

Build habits

If safe behaviour can be established as a custom and practice then desired behaviour may become more habitual and automatic, in the same way that we put on a seat-belt prior to driving a car, without really thinking about the activity.

A challenge is presented, however, in shifting from the temptation to take short-cuts in OSH, to a situation where safe behaviour becomes habitual. It should be recognised that building habits in this manner cannot be achieved overnight and will arguably require a gradualist approach to ensure success. Therefore, efforts should be exerted on influencing OSH outside of the workplace to encourage safe behaviours when people come to work. Migrant workers in some countries live in labour camps where OSH and welfare standards are often substandard. If organisations pay closer attention to improving OSH in the living environment, bolstered by regular inspections and audits, then this can be a way of developing an understanding that safety is a life skill and not just something that should be applied in the workplace.

Social consensus

Most individuals are conformists, with the concept of "social proof" indicating that we constantly observe others to verify our own behaviours (Cialdini, 2007). If behaving safely becomes the socially acceptable way of acting, then there is a greater chance of other employees adopting similar behavioural patterns. Creating an environment of "psychological safety" may also help workers in becoming more comfortable in raising OSH issues in the workplace (Edmondson, 1999).

One way to rally employees can be through the implementation of awards to workers who are seen to demonstrate excellence in OSH performance. For example, an "Employee Safety Award of the Month" program can be initiated, whereby safe behaviours of employees are celebrated. This process should be managed in a fair and consistent way, with awards given to employees who provide suggestions for improving OSH, or perhaps best practice in relation to OSH when performing a safety critical activity. A photograph of the worker could be posted on project notice boards to develop the recognition that safe behaviour is synonymous with success.

Over time, social pressure may ultimately create a critical mass and contagion in safe behaviours. OSH change efforts often require a hands-on approach and involvement from workers. A growing body of evidence indicates that attempts to actively engage workers in decision making can contribute to improvements in OSH performance (Fidderman & McDonnell, 2010).

Motivation

The literature on motivation and incentives is substantial and beyond the scope of this book. However, in consideration of the need to consider emotional, rational and situational factors in OSH, it is important to ensure that incentive schemes are designed carefully to facilitate the desired response. The problem in assigning incentives linked to numerical targets (e.g. number of reported accidents) is that individuals typically find a way of reaching the target, irrespective as to whether it is realistic (actual/truthful) or not. Under these conditions, the focus on performance is to the detriment of genuine learning. For example, in the case of accident reporting targets, financial incentives for meeting targets can lead to under-reporting, with limited recognition of associated learning points. Subsequently, behavioural change via incentives of this nature is often short term.

To change behaviour permanently there is a need to reach out to a person's identity or value system. In this sense, it is better to link incentives to core values as opposed to numerical targets. Incentives aligned to values will move the organisation from a performance based orientation to learning focused (Marquardt, 2011). A learning based approach will ensure that workers truly understand the implications of their actions and are more likely to engage in behaviours that will improve their OSH performance.

Human resources and OSH

HR is all about people and fundamentally OSH is no different. Subsequently, the HR department is arguably one of the most important allies in the journey to improving OSH culture and achieving excellence in OSH performance. However, far too often the relationship between OSH and HR functions can be soured by unclear goals. HR management is fundamentally a management responsibility; therefore, to a large extent, HR's mandate with respect to OSH is to support line management and the organisation by creating and overseeing policies, procedures and programs and advising, coaching and training line managers and employees.

 When working on OSH culture change programs remember to engage with your HR function as cross-over points will occur in many areas.

There are cross-over points between OSH and HR and it is important to adopt a strategic approach to exploring how a mutually beneficial relationship can be established. This should be established on objectives developed in unison, with clear roles and responsibilities and regular communication. An overview of the common areas where OSH and HR departments can collaborate is detailed below:

- Health and well-being – working together on employee wellness and engagement programs. These programs try to make employees more aware but also responsible and accountable for their own health and well-being
- Harassment and bullying – this is a serious issue in some workplaces and can be an underlying reason for stress which, as an occupational health issue, can lead to occupational ill health. Whilst the problems that arise can manifest themselves as OSH issues, their solution is generally through the implementation of effective HR policies and processes

- Attendance management, including absenteeism and presenteeism/return to work – these are recognised as major issues that impact directly on productivity, OSH performance and various long term organisational issues
- Disability management – it is critical that disabilities are recognised and managed at work. Employees need to be supported to ensure that they continue to be productive, feel valued and that work does not aggravate any disabilities. Physiological support is often necessary and these arrangements need to be incorporated into an organisational HR system
- Worker's compensation claims – whilst this is a legal matter, HR generally need to work closely with the OSH function in review panels and medical boards to assess cases and help in the effective management and apportioning of claims
- Job design – job design, including effective ergonomic and related OSH considerations should be performed between HR and OSH practitioners
- Training – this is very important as workers need to be aware of their OSH duties and responsibilities. Training provision, often starting with OSH induction training, will typically be a collaborative effort between HR and OSH functions.

 Mates in Mind

The Mates in Mind (Mates in Mind, 2017) campaign was established by the UK Health in Construction Leadership Group and the British Safety Council to raise awareness of mental health, help people understand how, when and where to get support and promote a positive culture of wellbeing in the construction industry.

According to the Centre for Mental Health (Sainsbury Centre, 2007), 91 million days are lost each year due to mental health problems. The Centre has calculated that the total cost to employers is estimated at nearly £26 billion each year – equivalent to £1,035 for every employee in the UK workforce.

Mates in Mind provides access to training and awareness, campaign materials and tailored support for your organisation. In addition, toolkits have been developed for workers where you can access a range of information and support tools.

Mates in Mind and other campaigns and initiatives focused on health and wellbeing are good opportunities for OSH and HR professionals to work closely together in addressing organisational issues. So, why not start today in trying to identify these types of opportunities to improve the relationship between OSH and HR.

Consultation and worker involvement

Another cross-over point between HR and OSH functions is consultation and worker involvement. This area is deemed of considerable importance and worthy of further discussion. It is good practice, and in many countries a legal requirement, to set up an OSH committee for consultation with representatives of employee safety elected by the workforce. The HSE has developed guidance on their website on the implementation of OSH committees (Health & Safety Executive, 2010). This guidance provides user-friendly advice for employers on how to set up an OSH committee and addresses the factors necessary to ensure continued success. The guidance details how the committee should work and recommends the creation of a written constitution detailing the basic rules and procedures for the operation of the committee.

OSH committees can often stagnate and lose the impetus they held when first created. Research has been carried out into this trend and a number of factors identified to help support the long-term effectiveness of an OSH committee (Health & Safety Executive, 2004). OSH committees should involve a variety of different people to demonstrate management commitment to differing concerns. Where committees are dominated by management, it is more difficult to establish a balanced perspective on OSH issues. Too often OSH committees become a complaints forum for raising trivial matters that receive no management attention. To avoid this, it is important that OSH committees address strategic issues affecting the workforce, or groups in the workforce, and allow day to day OSH issues to be resolved at a local level. Committee meetings should not be cancelled unless there are exceptional reasons, otherwise frequently cancelled, or postponed meetings, can damage the value of the committee.

A degree of formalisation is required to foster effective OSH committees. Meetings need to be planned, with all committee members receiving a personal copy of planned meeting dates. An agenda should also be provided prior to commencement of the meeting. During the meeting, minutes need to be taken with clear actions and responsibilities defined. This establishes accountability and allows the close out of actions to be addressed in future meetings. A level of etiquette should also be maintained during meetings. It is good practice to incorporate rules of conduct or behaviour in the committee's constitution, with members understanding the need to follow these conventions and conform to acceptable standards of behaviour. An assertive chair can help address disagreements and ensure the smooth running of the committee. Overall, consideration of these factors helps assist in maintaining the credibility and momentum of the OSH committee over a sustained period.

Performance management

Another area of collaboration between OSH and HR that requires more detailed discussion is performance management, the process of ensuring that individuals and teams contribute to high levels of organisational performance. It is essential

Worker involvement

HR and OSH functions can collaborate effectively in the design of worker involvement mechanisms to promote greater levels of engagement.

Consider your current approach to worker involvement in OSH. Write down a number of ways in which you could encourage greater worker involvement in OSH.

So, what did you write down? Fundamentally, there are opportunities to involve workers in OSH at every stage in the development, implementation and maintenance of an OSH management system.

- Policy – hold workshops with employees to agree on a vision and objectives that could be incorporated into the policy
- Planning – get workers involved in risk and opportunity assessments and the design of control measures that are derived from these assessments
- Implementation and Operation – involvement in the provision of OSH training or enrichment of job roles to incorporate additional OSH responsibilities
- Checking – employees can contribute to reactive approaches (i.e. incident investigations) and active ways (i.e. inspections and audits) to measure OSH performance
- Management Review – worker participation in OSH reviews can be encouraged at an individual, departmental, site, group and/or organisational level.

So, try to adopt a systematic approach to your worker involvement initiatives and ensure that close liaison with the HR function is maintained.

---

for everyone in an organisation to know what is expected from them and the criteria by which performance will be evaluated. This is often implemented in practice via some form of appraisal process. An effective appraisal should be:

- Related to the job description and scope – this is important as people should be measured against their agreed job role and remit
- Established in clear goals – objectives and targets can be created, or clearly defined qualitative descriptive measures
- Documented and agreed before the assessment period starts.

The frequency of appraisals should be clearly defined and in most organisations, are conducted on an annual basis with mid-year reviews. As organisations are dynamic there should be an agreed mechanism to modify and make changes within the appraisal process. OSH practitioners need to work closely with HR managers to help them and other managers in establishing OSH related personal targets along with departmental and operational targets for OSH performance.

Appraisals can be undertaken in a number of ways:

- Behaviourally Anchored Rating Scales (BARS) – where the performance of an individual in different job aspects is measured against specific defined behaviours
- Quantitative targets based on specific and verifiable objectives – where it is possible to measure an employee's performance against set targets
- Evaluation of performance against traits – usually undertaken with senior employees where leadership behaviours are evaluated. This method can be subjective with a preferred evaluation method for managers incorporating 360-degree feedback.

Performance management has important applications within OSH as it can be used to ensure that individuals are encouraged to behave in ways that will have a positive impact on OSH. There are a range of performance management responses to behaviour in responses to safe and unsafe practices that can be used to modify behaviour. These approaches are often incorporated in behaviour based approaches to OSH discussed previously.

With respect to shaping positive OSH behaviours, it is widely regarded that positive reinforcement is the most powerful. In positive reinforcement, the individual receives something that is wanted or valued after demonstrating proper behaviour, compared to negative reinforcement, which as the term suggests, involves the use of a reprimand or disciplinary action, to ensure that individuals do not partake in unwanted behaviour. Positive reinforcement is simple and economical to implement, often by giving a few words of encouragement following observance of good practice. However, positive reinforcement should be applied consistently throughout the workforce, ensuring that managers and supervisors avoid differentiating between different people and work practices.

Any approach to performance management in relation to OSH behaviours must be fair and transparent and should be the basis later for any performance related rewards, promotions, commendations or disciplinary action. If it is applied in a consistent manner it can have a powerful impact on behaviours and attitudes and help to foster better working relationships.

Summary

The management guru Peter Drucker famously said: "Organisational culture eats strategy for breakfast." Indeed, you can have nicely crafted strategies, policies and processes but without the unwavering commitment and ownership from the people in your organisation you will never achieve your OSH goals. Shaping

OSH behaviours is challenging, with no "one-size fits all" strategy. However, all individuals have an emotional and a rational side to their behaviour. The conflict between these two dimensions is a constant daily challenge that we are all faced with. The environment or situation we are presented with is also a critical factor influencing behaviour with the recognition that unsafe behaviour is often more attributable to the environment, than to individual failings. The emotional/rational dichotomy and the role of environmental factors is a valuable tool in the armoury of OSH practitioners, with consideration to be given to these factors in the design of behavioural interventions.

The work that OSH and HR functions undertake shares many overlaps and commonalities as fundamentally they are people-centric professions. Indeed, the development of a strategic alliance with your HR department will help in identifying collaborative opportunities, particularly in relation to health and well-being and broader implementation of behaviour change interventions. So, take some time to think about the areas where you can work together and what you can be doing to strengthen existing relationships to build a partnership that really does make sure that people go home in a safe and healthy condition every day.

Key takeaways

- Organisational culture relates to shared attitudes and beliefs about what is important in an organisation, with the OSH culture being a sub-set of the wider organisational culture – *are you paying enough attention to culture in your organisation?*
- The level of cultural maturity is important as it has a profound influence on the selection of interventions selected to improve OSH performance – *are you aware of the level of OSH cultural maturity in your organisation?*
- Providing your organisation has a mature OSH culture, the introduction of a behaviour based approach can be used to generate a step-change in OSH performance – *are emotional, rational and situational factors sufficiently addressed in your OSH behaviour change programs?*
- Worker involvement and consultation helps to generate ownership and commitment and is essential to the success of your OSH programs – *are you doing enough to actively involve employees in OSH?*
- There are multiple cross-over points between HR and OSH functions and it is important to adopt a strategic approach to identifying collaborative opportunities – *are you developing a meaningful alliance with your HR department?*

References

Cialdini, R. (2007). Social proof: Truths are us. In *Influence: The psychology of persuasion.* New York, NY: HarperCollins Publishers, 87–125.
Dickenson, D. L. (2001). The carrot vs. the stick in work team motivation. *Experimental Economics*, 4, 107–124.

Edmondson, A. (1999). Psychological safety and learning behavior in work teams. *Administrative Science Quarterly*, 44(2), 350–383.

Eisenhardt, K. M., & Sull, D. N. (2001). Strategy as simple rules. *Harvard Business Review*, 79(1), 106–116, 176.

Fidderman, H., & McDonnell, K. (2010). *Worker involvement in health and safety: What works?* HSE and Royal Society for the Prevention of Accidents Report (RoSPA). Edinburgh, UK: HSE Books. Retrieved from www.hse.gov.uk/involvement/rospa-wish.pdf.

Fleming, M. (2001). *Safety culture maturity model.* Sheffield, UK: HSE Books, Offshore Technology Report.

Gilovich, T., & Eibach, R. (2001). The fundamental attribution error where it really counts. *Psychological Enquiry*, 12(1), 23–26.

Grivna, M., Aw, T-C., El-Sadig, M., Loney, T., Sharif, A. A., Thomsen, J., Mauzi, M., & Abu-Zidan, F. M. (2012). The legal framework and initiatives for promoting safety in the United Arab Emirates. *International Journal of Injury Control and Safety Promotion*, 19(3), 278–289.

Health & Safety Executive (HSE). (1999). *Reducing error and influencing behavior HSG48.* Edinburgh, UK: HSE Books, 2nd edition. Retrieved from www.hseni.gov.uk/hsg_48_reducing_error_and_influencing_behaviour.pdf.

Health & Safety Executive (HSE). (2002). *Behavior modification to improve safety: Literature review* (Offshore technology report 2000/003). Edinburgh, UK: HSE Books. Retrieved from www.hse.gov.uk/research/otopdf/2000/oto00003.pdf.

Health & Safety Executive (HSE). (2004). *Workforce participation in the management of occupational safety and health.* HSE Report, HSL/2005/09.

Health and Safety Executive (HSE). (2010). *Health and safety committees.* Retrieved from www.hse.gov.uk/involvement/hscommittees.htm

Health & Safety Executive (HSE). (2002). *Safety culture: A review of the Literature.* Sheffield, UK: HSE Books, HSL/2002/25.

Heath, C., & Heath, D. (2010). *Switch: How to change things when change is hard.* New York, NY: Broadway Books.

Kahneman, D. (2011). *Thinking, fast and slow.* London: Allen Lane, Penguin Group.

Katzenbach, J. R. (2000). *Peak performance: Aligning the hearts and minds of your employees.* Boston, MA: Harvard Business School Press.

Loney, T., Cooling, R. F., & Aw, T-C. (2012). Lost in translation? Challenges and opportunities for raising health and safety awareness among a multinational workforce in the United Arab Emirates. *Safety and Health at Work*, 3(4), 298–304.

Marquardt, M. J. (2011). *Building the learning organization: Achieving strategic advantage through a commitment to learning.* Boston, MA: Nicholas Brealey Publishing, 3rd edition.

Mates in Mind. (2017). Retrieved from www.matesinmind.org/.

Michie, S., van Stralen, M. M., & West, R. (2011). The behaviour change wheel: A new method for characterising and designing behaviour change interventions. *Implementation Science*, 6, 42.

Roughton, J., & Mercurio, J. (2002). *Developing an effective safety culture: A leadership approach.* Woburn: Butterworth Heinemann.

Sainsbury Centre. (2007). *Mental health at work: Developing the business case.* London, UK: The Sainsbury Centre for Mental Health.

5   Economics and financial management

Welcome to Step 5 of the MBA programme. We are half way through the journey of transformation as an OSH practitioner. This Step of the programme is all to do with money, which as we all know makes the world go around! We will look at economics, the study of income, wealth and well-being and its relevance to the world of OSH along with financial management and the importance of understanding the role of the finance department and financial metrics in your organisation.

On completion of Step 5 you will be able to:

*   Identify the principles of economics and its relevance to OSH
*   Explain the areas of macroeconomics and microeconomics within the context of OSH
*   Recognise what is meant by the changing world of work and the OSH challenges created
*   Understand financial management and the importance of developing knowledge in finance
*   Use a range of financial management tools to help present OSH as an investment as opposed to a cost.

Economics

Economics is the study of factors that influence income, health and well-being. The economic perspective on OSH encompasses both causes and consequences; the role of economic factors in the causation of workplace incidents and ill health and the effects of OSH on the economic prospects on individuals, organisations, industries, nations and even the world. Economics explains how people interact with markets to get what they want or to accomplish certain goals. Fundamentally, economics is split into two disciplines, namely macroeconomics and microeconomics.

Microeconomics

Microeconomics is primarily concerned with single factors and the effects of individual decisions. This branch of economics considers the behaviour of individuals and firms in making decisions regarding the allocation of limited resources. In OSH management, risks and the dealing with risks at an organisational level is typically viewed from a micro-economic perspective that is based on information available at that time and taking economic issues into account. The following sections will provide an overview of key microeconomic theories and their relevance and application to OSH.

Supply and demand

Supply and demand is one of the fundamental pillars of microeconomics. Demand refers to how much (quantity) of a product or service is desired by buyers. The quantity demanded is the amount of a product people are willing to buy at a certain price; the relationship between price and quantity demanded is known as the demand relationship. Supply represents how much the market can offer. An interesting example of the principle of supply and demand, in the context of OSH, is compensating wage differentials, i.e. when people are paid a wage premium when performing risky jobs (Dorman & Hagstrom, 1998). It is known that "danger money" is paid in certain industries where the probability of being injured or possibly killed is high. Clearly, this is a supply and demand issue, in that there will be less people willing to accept the risks with this type of work resulting in the need for employers to offer higher salaries.

Elasticity

In economics, elasticity refers to the degree of responsiveness in supply or demand in relation to changes in price. If the curve is more elastic, then small changes in price will cause large changes in quantity consumed. If a curve is less elastic (inelastic), then it will take large changes in price to effect a change in quantity consumed.

If you are selling OSH products or services, it is important to be aware as to whether the demand is elastic or inelastic. For products or services that are easy to commoditise then demand is likely to be inelastic. For example, the demand for providing OSH management system development services in many jurisdictions is likely to be elastic. The reason for this is that there are often many organisations providing these types of services and cost can be the deciding factor when selecting an organisation.

Elasticity of demand emphasises the need to try and be innovative with respect to the OSH products and services that you sell so that customers are willing to pay a price premium. If you are providing a unique product or service that is difficult to replicate then demand is likely to be inelastic. For example, if you are providing unique OSH leadership or behavioural consulting services, you may find that there are a smaller number of organisations providing these services and able to charge a higher price. Alternatively, if an OSH product is difficult to replicate, possibly due to intellectual property rights, then again it is likely that demand would be inelastic.

Consumer demand theory

Consumer demand theory analyses consumer behaviour, based on the satisfaction of wants and needs generated from the consumption of a good. One of the most important lessons from consumer demand theory is the law of diminishing (marginal) returns, which states that at a certain point, an additional factor of production causes a relatively smaller increase in output. The concept of marginal safety utility is a good example of how the principles of consumer demand theory have extended into the field of OSH management (Genserik et al., 2016).

Safety utility essentially refers to the OSH benefits that are derived when an organisation spends more money on OSH measures. If an organisation invests more in OSH, it is assumed that OSH performance will improve. However, a point will be reached where the marginal safety utility, or additional benefit, derived from this expenditure decreases, or in other words, the improvement in OSH performance is not justified by the amount that needs to be spent. This understanding is central to the concept of "reasonably practicable" which involves weighing a risk against the trouble, time and money to control it.

Game theory

Game theory is concerned with predicting the outcome of games of strategy in which the participants (for example two or more businesses competing in a market) have incomplete information about the others' intentions. The classic example of game theory is the Prisoner's Dilemma, a situation where two prisoners are being questioned over their guilt or innocence of a crime (Neumann et al., 2007). They have a simple choice, either to confess to the crime (thereby implicating their accomplice) and accept the consequences, or to deny all involvement and hope that their partner does likewise.

As OSH, like economics, is a social science, game theory has been applied in various settings. Game-theoretic models have been created in the context of chemical engineering to assess situations, including investment decisions, to improve OSH prevention arrangements (Reniers & Pavlova, 2013). Game theory has also been applied in relation to road safety focusing on possible strategic interactions between road users in different situations and how this information can be used to improve safety conditions.

Behavioural economics

Historically, economists can be accused of having ignored behavioural issues. However, recent times have seen an upsurge in interest generated by the failure of conventional economics to adequately address recent economic reality. Several theories have been identified in research conducted by the HSE that could be relevant in OSH policy making, including: whether there is a skewed perception of risk; a cost of processing information; compliance with OSH that might be affected by the level of stakeholder involvement and/or employees' perceptions of fairness; that the act of publicly committing to standards affects OSH performance; and the monetising of non-compliance through fines can affect OSH outcomes (Health & Safety Executive, 2009a).

 Remember that OSH is fundamentally a social science, so an understanding of behavioural aspects will always be important in what we do.

Costs of production

In microeconomics, the costs of production relate to the different expenses that an organisation faces in producing a good or service. These can include the following:

- Fixed costs – costs which do not vary with output (e.g. costs of new premises)
- Variable costs – costs which do vary with output (e.g. raw material)
- Semi-variable costs – costs which exhibit both fixed and variable factors (e.g. employing workers during a slow-down in production)
- Short-run costs – costs associated with production due to productivity losses in the early stages of operating a new organisation
- Long-run costs – diseconomies of scale that could occur when increased output leads to higher long-run average costs. (N.B. The long-term ability to vary factors of production can also have a positive impact when realising economies of scale, when increased output can lead to lower long-run average costs).

To make profits, organisations need to control costs. Therefore, it is important to be aware of the costs associated with OSH management and how these are budgeted in an organisation. Further discussion on this topic is presented later in this Step of the program on financial management.

Competition and market structure

Competition and market structure considers the various market structures in which organisations can operate, how structure impacts organisational behaviour, efficiency and the level of profits that can be generated. The key market structures are as follows:

- Perfect competition – many organisations are present, freedom of entry, homogeneous product, normal profit
- Monopolistic competition – freedom of entry and exit, but organisations have differentiated products. Likelihood of normal profits in the long term

- Oligopoly – an industry dominated by a few organisations exhibiting interdependence
- Monopoly – one organisation dominates the market. Barriers to entry are evident and the existence of supernormal profit.

Market structure and dynamics is another micro-economic consideration that can impact OSH outcomes. For example, in highly competitive markets individual organisations may be more likely to try and avoid bearing OSH costs. However, in a monopoly situation where an organisation is generating supernormal profits the funds available for OSH investments may be greater.

Labour economics

Labour economics looks at the supplies of labour services (workers) and the demands of labour services (employers) and attempts to understand the resulting pattern of wages, employment and income. Labour economics is closely linked to OSH management, with interesting dimensions including the effectiveness of economic incentives in improving OSH outcomes. The subject of motivation and the use of incentives shaping positive OSH behaviours is well documented (see Step 4 on *Organisational Behaviour and Human Resource Management* for further discussion on this topic).

Welfare economics

Welfare economics is a branch of economics that uses microeconomic techniques to evaluate well-being at an economy-wide level. Application of welfare economics typically involves government intervention to provide social welfare. In the context of OSH, welfare economics provides the foundation for various instruments of public economics, such as cost-benefit analysis. Government agencies and departments are required to prepare impact assessments when proposing developments in OSH legislation, such as Regulatory Impact Assessments (RIAs) performed in the UK (Health & Safety Executive, 2014). It is good practice for economists to provide advice and support for this, including examining the rationale for government intervention, identifying potential options, and assessing costs, benefits and risks.

Macroeconomics

Whereas microeconomics looks at economics from a more granular level, macroeconomics relates to the wider aggregate economy. Economists assess the success of an economy's overall performance by considering a range of macroeconomic variables, including:

- Growth – economic growth is typically measured by Gross Domestic Product (GDP). GDP equals the total amount value of goods and services produced in a country during a year. Economic growth is important as it relates to increases in personal health, welfare and prosperity

- Unemployment – the socio-economic benefits of employment are well documented with the unemployment rate a key indicator of the condition of the labour market. A low employment rate is an indicator of good economic performance
- Inflation – this concept relates to changes in the overall level of prices. Inflation is an economic concern as when the inflation rate is high the real value of money erodes, making it difficult for people to keep up with the rising cost of living
- Trade – international trade considers the financial transactions made between consumers, businesses and governments in one country with others. International trade expands markets for organisations as they are not limited to trading within their own borders.

This overview of macroeconomic indicators is all very interesting – but what is the relevance to the OSH practitioner? Importantly, the attention given to OSH in any country can often be influenced by macroeconomic factors. For example, during recessionary periods (i.e. downturns in economic growth) there can be pressure on organisations to achieve cost efficiencies. Against these pressures, it is important that the OSH practitioner can present a sound business case for OSH. Furthermore, OSH management and practice continues to take place against a backdrop of changing factors at work.

 Understanding the context of your organisation, as required by ISO 45001, will require knowledge of external factors, including economic factors, to effectively implement an OSH management system.

The changing world of work

The message that the world of work is changing is not new, however, in recent years the influence of change to the economy and labour market appears to be growing. In Step 1 on *Strategy and Leadership*, the importance of horizon scanning was discussed and the need for OSH practitioners to be aware of external factors and their potential impact on OSH performance. From a macroeconomic perspective, some of the powerful forces that continue to shape the world of work and the associated implications for OSH management are detailed below.

Globalisation

Organisations are increasingly outsourcing operations and displacing workers to different parts of the world. This creates governance challenges, as organisations have operating companies located in different countries, with the employer owing all employees a duty of care and required to maintain a consistent and organised approach to OSH management. Globalisation requires organisations to develop an understanding of requirements for OSH management in different countries, along with increasing awareness of cultural issues that may influence OSH and how these issues should be managed.

Information and communication technologies

The workplace continues to see huge growth in the use of Information and Communication Technology (ICT), as well as rapid changes in the type of technology used, including computer networks, electronic data interchange and the internet. The growing emergence of some occupational health issues are linked to the influence of ICT. Ergonomic problems associated with the use of display screen equipment are a good example of how the hazard profile faced by organisations can alter due to the changing world of work.

Non-standard patterns of employment

Further to the changes in the economic landscape, there continues to be an increase in non-standard patterns of employment. The standard pattern of employment relates to employees directly employed by an employer, undertaking work activities as detailed in an employment contract. However, non-standard patterns of employment are becoming more common. These include an increased use of contractors, self-employed workers, part-time, temporary, teleworking arrangements and a growth in shift working. These non-standard forms of employment create potential challenges of OSH management:

- Contractorisation – this refers to the increased reliance on contract labour in industry. Contractors are regularly used for specialist functions, when competent staff are not always present internally, but also for a range of support functions, such as security, cleaning and catering. Contractors are often transient and peripatetic, undertaking work on a wide variety of different sites, which can make it difficult to engender levels of motivation and commitment towards OSH
- Self-employment – this relates to individuals who work for themselves, as opposed to another company or person. People who are self-employed may struggle in allocating sufficient time and resources for OSH management, in comparison to larger organisations
- Part-time working – work is generally considered to be part-time when employees are contracted to work for anything less than normal basic full hours. If employees are not working full-time, there can often be a management perception to be addressed that there is less of a requirement to involve these individuals in OSH matters
- Temporary working – a temporary worker is typically someone employed for a limited period whose job is usually expected by both sides to last for

only a short time. Temporary workers tend to be used when the number of workers required on a project vary, or to provide cover for permanent staff on holiday, or maternity leave. Temporary workers are often recruited through an employment agency so it is important that sufficient consideration is given to OSH during screening and selection processes

- Teleworking – this refers to where employees conduct all or part of their working week at a location remote from the employers' workplaces, with homeworking regarded as a form of teleworking. A specific risk factor to be considered for teleworking is lone working, as often people are working by themselves without close supervision

- Shiftworking – this typically relates to a pattern of work where one employee replaces another on the same job within a 24-hour period. Substandard practices are often prevalent during night shifts, due to lower levels of supervision and greater levels of worker fatigue.

 Non-standard patterns of employment are increasingly common, so make sure you have identified potential OSH issues associated with these different ways of working.

Demographics

A striking demographic change impacting many economies is an ageing workforce. For example, in the UK, the number of people over 60 is expected to rise 50% by 2030 (Health & Safety Executive, 2004). This presents new OSH challenges, as older workers are required to remain in employment and contribute to society. Older workers are likely to require flexible working arrangements and may also have a greater propensity for homeworking.

Another emerging trend in some countries is the increasing percentage of women in the workforce. In the UK, participation rates of women have risen quite steadily over the past 20 years. Many women work in the service sector, with the growth in the service sector likely to lead to more women being in paid employment. Although the same standard is applied for both women and men, in terms of work methods and work equipment, women may face different risks to men in the workplace, due to the fact that women are physically different from men and that women tend to work in specific areas, unlike men, who are evenly spread across all occupational groups.

Migrant labour

Another significant macroeconomic phenomenon linked to globalisation is the increasing use and reliance upon migrant labour. It is important to note that in many countries, labour standards associated with migrant labour are increasingly in the media spotlight and represent a potential significant reputational risk if not managed effectively. The HSE has undertaken research assessing the OSH risks associated with the influx of migrant labour in the UK. This research suggests that migrant workers experience higher levels of accidents and higher levels of physical and mental ill health that is work related (Health & Safety Executive, 2009b).

In the UK, the HSE has developed a specific migrant workers website (Health & Safety Executive, 2015) and improved its multi-lingual information to provide guidance to workers from overseas, catering for a diverse range of languages. In addition, further guidance has been developed by the Trades Union Congress (TUC) aimed at safety representatives and other union officials who work with migrant workers, to ensure that their rights are protected (Trades Union Congress, 2007). This guidance focuses on the importance of engaging migrant workers and gives practical advice and examples of how safety representatives can promote improved OSH practice. However, in many countries workers do not have the ability to join trade unions due to legal restrictions. Notwithstanding, where possible, it is important to establish arrangements to encourage worker involvement and consultation on OSH matters.

Health and Safety in a Changing World

The Institution of Occupational Safety and Health (IOSH) commissioned a five-year research programme, Health and Safety in a Changing World, to explore the landscape of OSH and its implications for developing solutions that provide effective protection for workers and their communities (IOSH, 2017).

Due to increasingly dynamic work environments, the research programme reports indicate that the future OSH professional will need to assist proactively with problem solving rather than mapping processes against rules, as dynamic environments require more emphasis on OSH outcomes, as opposed to the means of achieving them.

The future OSH professional is also more likely to be acting in a consulting or coaching capacity to workgroups or project teams, with a recognition of when complementary expertise is needed to address emerging problems.

In the context of the changing world of work and the shifting social, economic, political and technological landscape, are you taking time to reflect on your role and address the learning and development needs which will ensure you remain credible and relevant as an OSH practitioner?

Financial management

One of the key areas that many OSH practitioners need to strengthen understanding is financial management. Finance matters to all organisations as they all have to bring in money and spend it to do business. On the bringing-in side key questions include:

- How much of our money comes from the owners, or from sales or borrowing?
- Which products or services and regions earn the highest profits and which ones fail to perform?
- How long does it take to collect money that customers owe us?

On the spending side, key questions include:

- Are costs what they should be – are we spending the right amount of money on people and other physical assets?
- If we only invest in several opportunities for growth which ones generate the most value?
- If we increased our output by 20% would we generate 20% more money?

Your organisation's finance department produces financial statements, budgets and forecasts. It is important that you can understand this information as the language and concepts will often be central to many of the conversations that take place in organisations and are the foundation for key decisions.

Budgetary responsibility

Previous discussion throughout the book has indicated that those with a commercial or financial background tend to end up in top management positions. A dilemma is presented in terms of how you can strengthen your understanding of financial matters if you are not involved in the financial aspects of running a business. Well the short and simple answer, as discussed in Step 2 on *General Management*, is to get involved!

The absence of direct involvement in financial management does not prevent you from taking an interest in the financial metrics that your organisation uses to measure success. One of the key characteristics that many of the best leaders possess is a curiosity to learn more about their organisation, including matters that you may not be involved in on a day-to-day basis (see more information about curiosity and other key soft skills in Step 10 on *Interpersonal Skills*).

Budget holders

In your organisation, the budget holders are those people accountable for expenditure from, and income to, budgets. Each budget holder is responsible for the control of his/her budget and the general financial administration of his/her area of responsibility. It is important to be aware of who are the budget holders, as those holding Profit & Loss (P&L) responsibility in your organisation are the people who tend to have significant power in making decisions. If you do not hold P&L responsibility then you can still look for ways in which you can get exposure to financial matters. Simply, showing greater interest in financial management and speaking with friends and colleagues that work in finance is an important way of learning more about this subject.

OSH as an investment

OSH is commonly referred to as an overhead function within business, as OSH is recognised as a support function and ongoing expense associated with operating a business. This status of OSH as an overhead expense creates a problem for many OSH professionals as the function is seen as a cost and not something which adds value (i.e. generates profit) for the business. However, it is essential that OSH is posited as an investment and not an expense to generate further buy-in and support from top management.

Our OSH Practitioner Insight Survey indicated that 76% of respondents agreed or strongly agreed that an understanding of investment appraisal is essential and in a world increasingly driven by austerity and cost efficiency, it is essential to be able to present OSH in a way that can be regarded by top management as a sound investment. As an OSH practitioner, you may be confident that effective OSH management is good business sense, however, not everyone thinks the same.

 76% of respondents either "agreed", or "strongly agreed" that understanding investment appraisal is important in making a strong case for my organisation to invest in OSH.

OSH is competing for resources in the same fashion as any other part of your organisation. Importantly, organisations do not have unlimited resources and have to be careful in the decisions that are taken as to where to invest money. Put yourself in the shoes of your CEO; if you wanted to invest money you would do so in the investments which are going to generate the greatest return. Subsequently, if you are presenting a case for investment in OSH, it is important that you can present a considered and robust financial argument for any new initiatives, projects and campaigns.

The concept of Return on Investment (ROI) impacts so many of the decisions that organisations make, along with our own personal investment decisions. For example, if you were given $100 today, or $100 in one year's time, what would you rather have? The answer should be that you would want the money today – and it has nothing to do with inflation. Quite simply, if you had the money today, you could invest it, say in an investment with a 10% annual profit/interest rate, and that $100 dollars would be worth $110 dollars in one year's time!

When a manager needs to compare projects, and decide which ones to pursue, there are generally three options available: Net Present Value (NPV), Internal

Life Savings

In 2013, Institution of Occupational Safety and Health (IOSH) introduced the Life Savings campaign aimed at showing how good OSH saves lives and money (IOSH, 2013). The campaign included a number of resources, including case studies, action plans and other resources, including an incident costing calculator. The incident costing calculator captures a range of immediate costs associated with the incident, such as first aid treatment and costs for making the area safe, along with longer term costs, such as sanctions and penalties and reputation impacts.

Often the hidden or indirect costs associated with incidents, particularly serious incidents, are underestimated. Subsequently, it is worth considering the application of an incident costing calculator in your organisation to demonstrate the business benefits associated with OSH.

Rate of Return (IRR) and payback method (Gallo, 2014). In simple terms, NPV is the present value of expected future cashflows. This concept is linked to the time value of money. IRR is a metric used in capital budgeting measuring the profitability of potential investments. Internal rate of return is a discount rate that makes the NPV of all cash flows from a project equal to zero. The payback method is a simple approach used to calculate when you will make back the money you put in to an investment. However, it doesn't consider that the buying power of money today is greater than the buying power of the same amount of money in the future. This is one of the reasons why managers often prefer to use NPV when evaluating investments decisions.

Imagine that your organisation is considering two OSH investments options: rolling out an OSH training course for all members of the senior management team and introducing a vehicle speed and tracking system onto a small fleet of vehicles used by employees. The training program will cost a total of $20,000 and the driver monitoring system a total of $25,000. If the organisation can only afford one of these options how does it decide which one makes economic sense?

By working out the ROI the organisation will be able to evaluate how much money the investment will generate compared with how much it will cost. Before commencing an ROI analysis, it is important to understand the costs and benefits associated with the investment. This will allow you to understand the relative merits of each investment against the consequences.

ROI calculations typically involve the following steps:

• Identify all the costs associated with the investment
• Estimate the savings to be achieved

- Determine how much money the investment will create
- Produce a time-line for expected costs, savings and cash flows, and use sensitivity analysis to challenge your assumptions
- Evaluate the unquantifiable costs and benefits.

For the two OSH investments presented above the costs will be relatively easy to determine. For example, costs associated with the training program may relate to design and development, administration, materials, facilities and labour associated with delivery. However, the challenge is in estimating the savings and determining how much money the investment will create. When OSH investments are made often the benefits are difficult to quantify, however, it is important that all benefits are captured and that attempts are made to assign a financial value to these benefits.

OSH: a cost or investment?

One of the key challenges facing OSH is that it is often regarded as a cost and not an investment. For example, in 2012, the UK Prime Minister at the time, David Cameron, quoted that OSH laws were holding back businesses. His view was that the cost that UK businesses had to face to ensure compliance with OSH laws impacted their competitiveness and impaired international trade.

During your career, you will encounter people that see OSH only as a cost to your business. Your success as an OSH practitioner will often depend significantly on your ability to influence these individuals regarding the business benefits of OSH.

Spend some time thinking about how you could present OSH as an investment as opposed to a cost and what else could be done to generate interest in OSH programs and initiatives.

Summary

An understanding of economics and finance is critically important to your career progression towards senior OSH roles and to ensure that you and the ideas that you present are perceived as credible. We live in a world that revolves around money and often the decisions that are made by organisations are influenced by short-term expectations. Try to ensure that you are attuned to the financial pressures of running a business and that OSH is positioned in a way that addresses these realities.

As an OSH practitioner, it is important to understand financial concepts so that you can ask essential questions and contribute to sensible decisions. So, don't complain that your CFO or finance team isn't interested in OSH; take time to understand their world so that you fully appreciate the critical financial drivers faced by many organisations. This may force you out of your comfort zone, however, the impact that it will have will be worth the effort.

Key takeaways

- Economics is a social science that considers the behaviours of individuals, groups and organisations regarding income and well-being – *have you taken time to understand the relevance of economics to OSH?*
- Macroeconomics considers the output of a nation as a whole whereas microeconomics looks at issues at the level of individual people and firms – *are you aware of the macro and microeconomic factors influencing OSH?*
- Macroeconomics forces can create significant impacts on OSH standards – *are you aware of the OSH risks and opportunities created by the changing world of work?*
- A knowledge of financial management is important for career progression to senior OSH roles – *do you possess knowledge of the financial metrics used to measure performance in your organisation?*
- The ability to use a range of tools to present OSH as an investment will have a significant impact on how OSH is perceived in your organisation – *are you familiar with investment appraisal, net present value, payback analysis and cost-benefit analysis?*

References

Dorman, P., & Hagstrom. (1998). Wage compensation for dangerous work revisited. *ILR Review*, 52(1), 116–135.

Gallo, A. (2014). *A refresher on net present value.* Cambridge, MA: Harvard Business Review. Retrieved from https://hbr.org/2014/11/a-refresher-on-net-present-value.

Genserik, L. L., Reniers, H. R., & Noel, V. E. (2016). *Operational safety economics: A practical approach focused on the chemical and process industries.* Hoboken, NJ: John Wiley & Sons.

Health & Safety Executive (HSE). (2004). *Thirty years on and looking forward: The development and future of the health and safety system in Great Britain.* Sheffield, UK: HSE Books, C25.

Health & Safety Executive (HSE). (2009a). *Behavioural economics: A review of the literature and proposals for further research in the context of workplace health and safety.* Sheffield, UK: Research Report 752.

Health & Safety Executive (HSE). (2009b). *Migrant workers in England and Wales: An assessment of migrant worker health and safety risk.* Sheffield, UK: Research Report 752.

Health & Safety Executive (HSE). (2014). *Impact assessments.* Retrieved from www.hse.gov.uk/economics/ria.htm.

Health & Safety Executive (HSE). (2015). *Migrant workers.* Retrieved from www.OSH.gov.uk/migrantworkers/.

Institution of Occupational Safety and Health (IOSH). (2013). *Life savings.* Retrieved from www.iosh.co.uk/About-us/What-we-are-up-to/Life-savings.aspx.

Institution of Occupational Safety and Health (IOSH). (2017). *Health and safety in a changing world*. London: Routledge (Edited by Robert Dingwall and Shelley Frost).

Neumann, J. V., Morgenstern, O., & Kuhn, H. W. (2007). *Theory of games and economic behavior*. Princeton, NJ: Princeton University Press, 60th anniversary commemorative edition.

Reniers, G., & Pavlova, Y. (2013). *Using game theory to improve safety within chemical industrial parks*. London: Springer-Verlag.

Trades Union Congress. (2007). *Safety & migrant workers: A practical guide for safety representatives*. London: TUC Publications.

6 Operations management

Welcome to Step 6 on Operations Management! Operations are the processes by which businesses create and deliver value. In this Step, strategic aspects of operations management are discussed, with a focus on the design and implementation of business processes related to the different organisational ways of working. It is intended that a more thorough understanding of operations and operations management tools will help the OSH practitioner ensure that OSH is deeply ingrained into the fabric of operations.

On completion of Step 6, you will be able to:

- Explain Operations and the role of the Operations Manager
- Understand what is meant by the value chain and the importance of embedding OSH into ways of working
- Determine the need to codify ways of working within business management systems and the benefits and limitations of integrated management systems
- Identify the ways in which OSH can be integrated into the different stages of the project life-cycle
- Outline a range of quality and operations management tools and understand their relevance to OSH management.

Operations

Operations is an incredibly broad term and means different things to different people. However, operations are typically associated with producing things that add value to a business. The focus here will invariably be on implementation. Understanding the effectiveness and efficiency of operations is fundamental to the successes of almost any organisation and industry. This is of specific importance to the manufacturing, processing and industries that have physical operations, such as transportation, logistics, aviation, oil and gas, engineering, construction, fabrication, etc.

Operations management is the activity of designing and managing processes to achieve results of value to the various stakeholders of an enterprise (Stevenson, 2014). Fundamentally, this will entail the implementation of a set of coordinated activities relying on various resources to transform inputs into outputs. The effective management of operations can lead to various benefits:

- Reduced waste from operations
- Improved efficiency and reduced operating costs
- Enhanced OSH performance through incident prevention
- Improved inventory management leading to better space and costs utilisation.

 Adopting a holistic approach

Operations management focuses on the transformation of input into outputs. From an OSH management perspective, it is important that a holistic approach is adopted to ensure that the OSH risks and opportunities associated with operations are managed.

Spend time considering your organisation and the various inputs and outputs that arise from the transformation process.

What are the significant risk and opportunities?

How can you ensure that OSH is embedded into the different aspects of operations?

Operations management is a key component of operational excellence which combines OSH and other related disciplines, such as environmental and quality management. Understanding operations is important as this knowledge will help the OSH practitioner contribute effectively to making operations safer, more reliable and operationally efficient. The ability to take a holistic view of operations

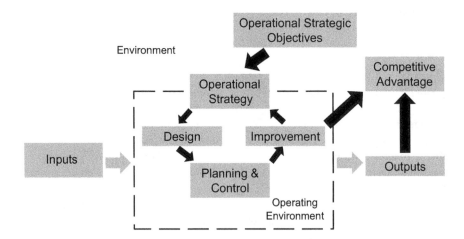

Figure 6.1 Operations management framework

will also enable the OSH practitioner to identify significant risks and opportunities and prioritise activities accordingly.

The role of the operations function

The role of the operations function varies across organisations; however, it can typically be defined by three key functions (Slack et al., 2004):

1 Implementer of business strategy – Putting strategy into action is called "operationalisation" of the strategy, involving activation of strategic imperatives through organisational functions. The operations function will lead on design and implementation of operational plans to enable execution of strategy. Whereas the CEO will be responsible for defining business strategy, it is the responsibility of the Chief Operations Officer (COO) to ensure that strategy is implemented.
2 Supporter of business strategy – Operations need to support strategy in terms of the development of capabilities and capacities, which in turn helps the organisation improve and refine its strategic goals. In this sense, operations management supports business strategy by ensuring that support functions operate in a coordinated manner to facilitate effective and efficient operational delivery.
3 Driver of business strategy – The most difficult and challenging role of the operations function is to drive strategy. In many organisations where strong relationships are required with suppliers, customers and other stakeholders, a substantial aspect of the value proposition of an organisation is linked to operations which in turn helps to drive business strategy.

OSH and the value chain

One of the common examples of management jargon referred to in MBA programs is the value chain. This term is important within operations management and needs to be understood by OSH practitioners. The value chain is all aspects of an organisation that come together to deliver value. A conceptual example of a typical value chain for a manufacturing and distribution company is presented in Figure 6.2.

The bottom section of the value chain depicts the logical flow of business functions that ultimately result in the provision of a service. The top section of the value chain shows the business functions that are in place to support service delivery, such as procurement, HR and information technology. These functions are commonly referred to as overhead functions as they are not profit making, but do play a critical role in supporting operations. Ensuring the right balance between operational functions and business support functions is a critical strategic and financial consideration, as top-heavy business support can impact profitability; however, insufficient business support can undermine the effectiveness of product or service delivery.

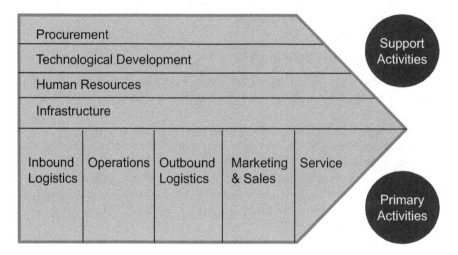

Figure 6.2 Example of a value chain

The operations of business functions, in terms of the ways of working, will typically be defined via some form of Business Management System (BMS), incorporating policies, processes and associated forms and templates. A key question for the OSH practitioner is the extent to which OSH is embedded into your value chain? Indeed, the need for OSH practitioners to be cross-disciplinary in their approach and looking to integrate OSH within business process is the theme within this second part of the book. The importance of a cross-functional approach to OSH management is broached in considerable detail in the new revision of 14001 and the recent publication of 45001 (see Step 2 on *General Management* for further discussion on ISO 45001).

Business process reengineering

A BMS will incorporate processes, which are in effect your organisational ways of working. Business Process Reengineering (BPR) is the design and analysis of work-flows within and between organisations to optimise processes and add value, be that improvements in productivity, quality, safety, etc. Operations Managers manage processes; in effect, day-to-day operations management centres on ensuring business processes are developed, implemented and maintained, along with identifying opportunities for continual improvement. Subsequently, for OSH practitioners to understand operations management it is important to have knowledge of BPR.

Process mapping

Process mapping involves the description of processes in terms of how activities within a process relate to one another. There are many techniques which can be used for process mapping, including process blueprinting, process analysis and process critical path analysis. Process mapping techniques typically incorporate two key features:

(a) Identification of different types of activities that take place in a process
(b) Depictions of the flow of resources or information/data through this process.

When studying complex processes, such as those present in major hazard industries, process flow diagrams and process and instrumentation diagrams will be used. When designing process flows it is important to consider the interactions and interdependencies that may exist between various processes. For example, when designing processes for procurement, invariably there will be opportunities to embed OSH and environment into procurement practices, which requires process owners to work closely together to create a truly cross-functional system.

Integration of management systems

The decision whether to integrate or maintain separate management systems is a question that many OSH practitioners will face during their career, a decision that has substantial relevance to operations management. The integration of OSH

Table 6.1 Integrated vs. separate management systems

Integrated management systems	Separate management systems
Consistency/avoidance of duplication and unnecessary cost in procedural, record-keeping, auditing and software areas.	Existing systems may work well and may lose some effectiveness during and after integration.
Avoids narrow decision making that solves a problem in one area but creates a problem in another area.	OSH and quality systems cannot be treated the same, e.g. OSH standards must meet legal minimal, and quality standards can be set internally. Separate systems can be operated more easily with a different philosophy.
Encourages priorities and resource utilisation that reflects the overall needs of the organisation rather than individual disciplines.	Business needs may demand systems of different complexity, e.g. a complex OSH system and a simple quality system, so bureaucracy can be more easily tailored to the needs of the subject.
Initiatives in one area may benefit other areas.	Separate systems may encourage a more detailed and focused approach to auditing and standards.
Encourages closer working and equal influence amongst specialists/career development.	Separate systems are clearer for regulators to work with.
Provides scope for the integration of other areas, e.g. security, production safety, etc.	Change to integration will require resource input/time to develop.
Encourages the spread of a positive culture across three disciplines.	

Annex SL
High level structure clauses
  1 – Scope
  2 – Normative references
  3 – Terms and conditions
  4 – Context of the organisation
  5 – Leadership
  6 – Planning
  7 – Support
  8 – Operation
  9 – Performance evaluation
  10 – Improvement

Figure 6.3 Annex SL high level structure

with quality management and environmental protection is a commonly applied business strategy, and, if properly resourced and managed, can help in achieving business efficiency and the associated benefits.

Integrated management systems tend to be more effective in large, complex, multi-hazard organisations. They are less beneficial for other types of organisations, particularly, when the degree of risk posed by their operations, to either OSH or to the environment, is low. For these organisations, the costs of introduction may be disproportionate to the benefits. Table 6.1 details the respective benefits of integrated and separate systems.

The design of management system specifications is heading in a direction to enable integration. Recent revisions of ISO 9001 and ISO 14001 have been developed in line with ISO Annex SL and ISO 45001 has also been created in line with this framework. An overview of the key clauses within ISO Annex SL are provided in Figure 6.3.

ISO Annex SL introduces common terminology applied across related management systems (i.e. ISO 9001, ISO 14001 and ISO 45001) and fosters effective integration of management systems (ISO, 2015). Several new concepts that are introduced by this high-level structure are discussed in Step 2 of the MBA program on *General Management*, including requirements to determine the 'Context of the organisation' and provisions for 'Leadership'.

Project life cycles

In project led organisations, operations will typically relate to the delivery of projects. A project is commonly defined as a unique, transient endeavour, undertaken to achieve planned objectives, which could be defined in terms of outputs, outcomes or benefits (Association for Project Management, n.d.). OSH management on projects is outside of the scope of this publication, however, it should be noted that the integration of OSH within project life cycles is just as important as it is

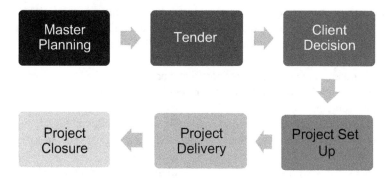

Figure 6.4 Example of a project life cycle

within business value chains. A project will entail various phases, as shown in Figure 6.4.

It is important for OSH practitioners to look for opportunities to embed OSH into all stages of a project. Invariably, the greatest yields will be gained by ensuring that OSH is considered during the planning and design phase of projects. For example, in the construction industry, the importance of safety in design is well-established, ensuring that hazards are eliminated where possible and information is provided on remaining risks that may become apparent at later stages in the life cycle of the project.

By working closely with planners, designers, architects, engineers, procurement specialists and other people involved in the up-front aspects of projects, the OSH practitioner is more likely to be able to create meaningful change. Silos can be created in projects that prevent these interactions from taking place; however, it is down to you to break down the walls! All too often, OSH practitioners get involved in the construction or operational phase of projects and find themselves firefighting with problems that could have been addressed at an earlier stage of the project.

 If OSH requirements are clear in contractual requirements this provides an opportunity for organisations to make a financial provision when tendering for projects.

Supply chain management

In today's global economy, businesses increasingly rely on outsourcing parts of their activities and processes. The key to managing OSH effectively throughout the supply chain is close cooperation between suppliers, manufacturers, retailers and any other organisations involved. Issues which can affect parties upstream or downstream need to be discussed freely and action taken to minimise adverse effect on others. There are a range of initiatives that can be implemented to drive OSH improvements through the supply chain, including:

- Set and enforce consistent, proportionate and realistic OSH standards
- Ensure suppliers of products and services provide evidence of effective OSH management
- Request that suppliers have membership of prequalification schemes and hold third party accreditation of recognised international standards, e.g. ISO 45001
- Work closely with suppliers to improve their OSH performance to help establish long term relationships based upon mutual trust
- Share good practice between organisations within the supply chain
- Recognise and reward suppliers for good and improving OSH performance.

Outsourcing

Outsourcing essentially means that an external organisation performs part of an organisation's function or process. Examples can include the use of off-shoring centres in different countries to perform business support functions, such as finance and information technology, along with the use of contractors to perform activities when a competency does not exist within the organisation. Some of the benefits associated with outsourcing include:

- Reducing operating costs
- Increasing manufacturing productivity and flexibility
- Leveraging the expertise and innovation of specialised firms
- Encouraging use of best demonstrated practices for internal activities.

If business functions or processes are outsourced to another organisation, although the external organisation will not be within the scope of your OSH management system, the functions and processes that are created are within the scope of your system. Subsequently, it is important that OSH requirements are communicated to third parties and that there are arrangements for collaboration to ensure that OSH is integrated within related functions and processes.

Contractor management

Contractors are regularly used for specialist functions, when competent staff are not always present internally, but also for a range of support functions, including security, cleaning and catering. In the most extreme cases, in some industrial environments, it is not uncommon to find up to half of the workforce of several thousand being made up of contract staff, carrying out a multitude of activities.

The hidden risks of outsourcing

On 24 April 2014, the Rana Plaza building in Dhaka collapsed, killing more than 1,100 people. The investigation determined that failures in design and construction of the building led to the incident, including problems with building materials. This tragic incident was a stark reminder of the need to safeguard vulnerable people in supply chains around the world.

Following on from the disaster there have been a number of initiatives, including developments in legislation and best practice, to prevent similar tragedies; however, significant further work is required to secure better working conditions in supply chains.

Understanding risk exposures across your supply chain is a fundamental aspect of risk management. Importantly, make sure to take time to assess your supply chain, with the understanding that not every country is at the same level of OSH development, but clear and consistent communication of OSH standards and expectations is important, supported by effective assurance programmes.

These contractors are often highly transient and peripatetic, undertaking work on a wide variety of different sites, which can make it difficult to engender levels of motivation and commitment towards the client. Generating effective OSH standards in a workplace that is made up primarily of contractors can present a substantial challenge.

Once contractors are on site, the use of company and contractor representatives can be a useful way of ensuring effective collaboration between the different parties. The company representative acts as a focal point for OSH issues between the company and the contractor. This includes ensuring that the contractor is aware of site standards and procedures and periodically checks that the contractor is meeting OSH obligations. Another way in which cooperation on OSH can be facilitated is to allow the contractor representative to attend OSH committee meetings. This provides an opportunity for the company to discuss relevant OSH issues with the contracting company, facilitating input from the contractor representative to address associated problems.

Quality management

The subject of Operations Management is invariably linked to Quality Management, largely due to its focus on effectiveness and efficiency. Total Quality Management (TQM) is a management approach that originated in the 1950's and steadily became more popular since the early 1980's (George & Weimerskirch, 1998). Fundamentally, total quality is a description of the culture, standards and

Contractor management

Guidance on contractor management produced by the HSE makes specific reference to the importance of co-operation and co-ordination in any client/contractor relationship (Health & Safety Executive, 2002). This guidance emphasises the need for organisations to implement a pre-qualification assessment when selecting contractors. This assessment can be used to determine whether the contractor has a suitable OSH management system in place.

Try to identify a range of other ways of promoting co-operation and co-ordination with contractors in the development, implementation and management of your OSH management system.

processes within an organisation that strives to provide customers with products and services that satisfy their needs. Invariably, a total quality management approach will run hand-in-hand with operations management as quality, like OSH, will need to be integrated into all aspects of operations.

The TQM philosophy stresses the importance of actively involving employees in quality management processes. It also recognises the crucial importance of visible leadership and the need for consistent emphasis on quality improvement throughout the organisation. In this sense, quality management shares many commonalities with OSH, in addition to the fact that TQM systems are based upon the "Plan, Do, Check, Act" model, commonly applied within OSH management. There are a number of tools that are applied within quality management that are worth spending some time discussing in the following sections.

If your organisation fails to devote enough time to the quality of its product or services then in many cases OSH risks and issues will arise.

Six Sigma

Six Sigma is a disciplined, data-driven approach and methodology for eliminating defects (driving towards six standard deviations between the mean and the nearest specification limit) in any process – from manufacturing to transactional and from product to service. The objective of Six Sigma is to improve the quality of processes by identifying and removing the causes of defects that could lead to quality defects. Although not often used in the OSH arena to full potential, Six Sigma has the power to significantly influence OSH standards by identifying and addressing unsafe working conditions and behaviours.

One area of overlap between OSH and Six Sigma is in relation to ergonomic issues. Insufficient consideration of the interface between people and their environment, the equipment they use and the task they are performing can result in OSH and quality concerns. Indeed, many organisations have used the Six Sigma methodology to encourage a systematic and objective approach for the management of OSH risks (Granger, 2012).

Lean manufacturing

Lean manufacturing is a quality management approach based on finding efficiencies and removing wasteful steps that do not add value to a product. Historically, lean manufacturing can be traced back to Henry Ford and the concept of continuous flow on assembly lines, whereby production standards were kept extremely light to avoid waste (Lean Enterprise Institute, n.d.). When considering waste, the following categories can be monitored:

- Overproduction – are you producing more than consumers demand?
- Waiting – how much lag time is there between production steps?
- Inventory (work in progress) – are supply levels and work in progress inventories too high?
- Transportation – do you move materials efficiently?
- Over-processing – do you work on the product too many times, or otherwise work inefficiently?
- Motion – do people and equipment move between tasks efficiently?
- Defects – how much time do you spend finding and fixing production mistakes?
- Workforce – do you use workers efficiently?

To get lean there could be the potential for OSH to be undermined resulting in increased OSH risks. However, if OSH can be embedded into lean manufacturing there is the opportunity to add additional value by ensuring that processes are better, less wasteful and faster. Indeed, with a carefully designed and implemented approach there is no reason as to why lean methodologies and associated objectives should result in situations where OSH is compromised.

 71% of respondents either "agreed", or "strongly agreed" that I have a good knowledge of the operations management tools applicable to my organisation.

Planning and control

Planning and control is important to operations management in ensuring that operational processes remain effective and efficient in producing set outcomes which the processes are designed to deliver. Planning involves the formalisation of what is intended to happen, and whilst the actual operations may not always run exactly to the plan, without setting planning targets it is impossible to assess the effectiveness of operations. Planning also helps ensure that the right amount of resources is identified and allocated.

Control is the process of managing changes and deviations from the plan. This often requires re-planning in the short term and interventions at various levels. In certain areas changes are dealt with directly by an operations team, in other areas where the deviation from the plan is significant, supervisory and management input may be required. OSH is important in this process and risk assessments need to identify and manage OSH related risks. Deviations should be studied in a systematic way to allow for any changes to take place within tolerance levels guided by appropriate safety margins.

In the major hazard industries, there are various tools used to ensure that OSH considerations are considered during the planning stage of capital expenditure. For example, Hazard and Operability Study, or HAZOP, is a common practice in the process industry, involving the work of a multi-disciplinary team considering the design intent of a new plant, for example, and assessing the various deviations away from design intent, with a view to proposing improvements to operability and safety at the design stage (Nolan, 2004).

Planning becomes complicated when there are uncertainties in relation to supplies, raw material variability and demand changes. One of the ways that efficiency in a system is improved is by regularly reviewing the latest estimates and setting short-term operational targets and goals to ensure effective resource utilisation. Manpower planning is another relevant example, as for example, on a construction site or during a shut-down of a process plant, the planning of manpower required daily is critical and can have a significant impact on costs and the risk exposure to people and assets. These are the type of challenges that will typically be dealt with by the Operations Manager, but should involve input from OSH practitioners.

It is beyond the scope of this book, however, OSH practitioners, particularly those working in processing and manufacturing industries should be aware of the need for planning and control, including the balance between capacity utilisation and demand. This is an area which is important as at times sudden increases in customer requirements and the opportunity to make higher profits by increasing production may introduce OSH risks. This requires effective management of change processes and a response that ensures that high OSH standards are maintained during this dynamic period.

Inventory planning and control is also an area that may be relevant to OSH practitioners, as at times market conditions may require stock generation, running a risk of over stocking, which can create OSH implications, particularly if stocks start to exceed the safe storage capacity of these materials. This may become even more of an issue in industries where dangerous materials, such as combustible, flammables and hazardous substances, are stored.

Summary

The role of the OSH practitioner invariably involves providing support to Operations Managers and others involved in the development, implementation and maintenance of business processes. It is fundamental that OSH practitioners possess a sound understanding of operations and related processes to aid effective decision making in this area. Importantly, operations management, quality management and OSH management go hand in hand as ultimately, they are all focused on delivering more effective and efficient activities.

OSH practitioners need to approach operations management with the aim of better understanding the OSH risks and opportunities of operations to provide effective support and guidance to prevent incidents and avoid major losses. The decision whether to integrate management systems or to retain separate systems will invariably be a key decision in many organisations. To develop a more cross-functional approach to OSH management, it is advisable to look for ways to integrate OSH into all aspects of operations, which in many cases can be achieved more effectively through the implementation of an integrated management system.

Key takeaways

- Operations is typically associated with producing things that add value to a business – *are you ensuring that OSH is considered during all aspects of operations?*
- The value chain is all aspects of an organisation, including delivery and overhead functions, that come together to deliver value – *is OSH effectively embedded into all aspects of your organisation's value chain?*
- There are various cross-over points between risk-related disciplines, with many organisations codifying ways of working through an integrated management system – *have you fully evaluated the benefits and limitations of separate or integrated systems?*
- For many organisations operations involves the delivery of projects, with projects typically entailing several key phases or stages – *is OSH integrated into all aspects of project management?*
- There are a range of approaches and tools used in quality and operations management which are applicable to OSH – *do you have an awareness of key operations management tools and their relevance to OSH practice?*

References

Association for Project Management (APM). (n.d.). *What is project management*. Retrieved from www.apm.org.uk/WhatIsPM.

George, S., & Weimerskirch, A. (1998). *Total quality management: Strategies and techniques proven at today's most successful companies*. Hoboken, NJ: John Wiley & Sons.

Granger, T. (2012). How Six Sigma can improve your safety performance. *Incident Prevention Magazine*. Retrieved from http://incident-prevention.com/ip-articles/how-six-sigma-can-improve-your-safety-performance.

Health & Safety Executive (HSE). (2002). *Use of contractors: A joint responsibility*. Sheffield, UK: HSE Books, INDG 388.

International Organization for Standardization (IOS). (2015). *ISO/IEC Directives, part 1 consolidated ISO supplement – procedures specific to ISO*, 6th edition. The Table of Contents taken from ISO/IEC Directives Part 1 Consolidated ISO Supplement – Procedures specific to ISO, is reproduced with the permission of the International Organization for Standardization, ISO. This document can be obtained from any ISO member and from the Web site of the ISO Central Secretariat at the following address: www.iso.org. Copyright remains with ISO.

Lean Enterprise Institute (LEI). (n.d.). *A brief history of lean*. Retrieved from www.lean.org/WhatsLean/History.cfm.

Nolan, D. (2004). *Applications of HAZOP and what-if safety reviews to the petroleum, petrochemical and chemical industries*. William Andrews. Saddle River, NJ: Noyes Publications.

Slack, N., Chambers, S., & Johnson, R. (2004). *Operations management*. Essex, UK: Prentice Hall, FT, 4th edition.

Stevenson, W. J. (2014). *Operations management: McGraw-Hill series in operations and decision sciences*. Columbus, OH: McGraw-Hill Education, 12th edition.

7 Marketing and brand management

In this Step, we will be looking at the function of marketing and how as an OSH practitioner you can develop your skills as a marketer. This is the final step that relates to the second theme in the OSH Practitioner Transformation Model on the need to "be cross-functional". A substantial amount of the content in this module relates to marketing of OSH products and/or services to external clients. However, many of the lessons herein can also be applied within an organisation when trying to develop levels of awareness and interest in OSH. Step 7 also includes content on developing and launching an OSH campaign within the context of existing organisational brand architecture and the importance of marketing yourself and being able to communicate your own personal brand.

On completion of Step 7, you will be able to:

- Identify the fundamentals of marketing and the importance of creating a marketing strategy
- Explain market analysis and the "5 C's" – Customers, Company, Competitors, Collaborators and Context
- Recognise the creation of a value proposition and market segmentation; targeting; positioning and capturing value, including the "4 P's" – Product, Place, Promotion and Price
- Understand the importance of marketing in developing organisational awareness and interest in OSH
- Appreciate the importance of marketing yourself and be able to deliver a personal pitch.

What is marketing?

The marketing function is the engine of any successful business. You can have the most talented people and offer the best product or service, but unless you have the capability to get people to buy what you are offering then your business will not survive very long! Thus, financial success often depends on marketing ability (Kumar, 2004). The marketing function will play a key role in identifying and meeting human and social needs, however, no amount of education and experience can create the creativity, empathy and intuition needed to be a great marketeer.

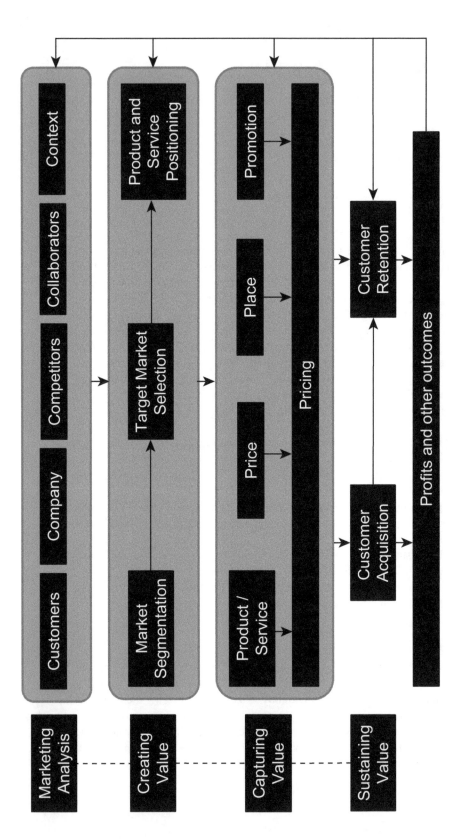

Figure 7.1 Marketing strategy framework

29% of respondents either "disagreed", or "strongly disagreed" that the marketing function in my organisation plays an important role for OSH.

Our OSH Practitioner Insight Survey identified a relatively high proportion (compared to other survey questions) of OSH practitioners felt that the marketing department in their organisation did not play an important role for OSH. This indicates that there is work to be done in building relationships with colleagues who work in marketing and leveraging the power of marketing for the benefit of OSH.

There is often a misconception that marketing is solely concerned with advertising of goods and/or services. However, marketing is a strategic activity with promotions only a small part of an effective marketing strategy. Marketing encompasses various processes and tools for creating value for customers, partners and society at large, which are detailed in the following sections.

Developing a marketing strategy

If you are working for an organisation that provides OSH products or services to clients it is important that you have a marketing strategy. An effective strategic marketing framework for the commercialisation of OSH solutions should consider the following, as shown in the conceptual model below (Dolan, 2000):

- Marketing analysis
- Creating value
- Capturing value
- Sustaining value.

Marketing analysis

The starting point in developing a marketing strategy for OSH products or services is to perform a marketing analysis. This should incorporate a study of the various facets of your business to gauge its current position and capability with respect to the intended product or service offering. A useful framework for ensuring consideration of internal and external factors (Chernev, 2014), is to address the "5 C's" – Customers, Company, Competitors, Collaborators and Context (just like OSH professionals, business scholars also love acronyms!).

Customers

The first step of the marketing analysis is to perform an assessment of your customers. This is a critical process and in isolation can represent a time-consuming and complicated process. However, if you do not understand your customers then

Theory – SWOT analysis

A method that can be used when performing a company analysis is SWOT analysis (Seth, 2015). SWOT examines the Strengths and Weaknesses of a company (internal environment), as well as the Opportunities and Threats within the market (external environment).

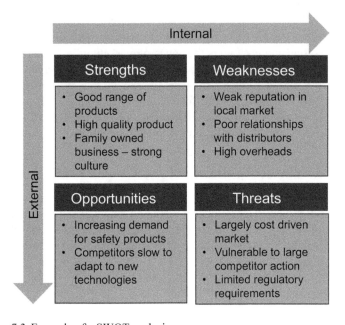

Figure 7.2 Example of a SWOT analysis

The example presented is a SWOT analysis for a family owned business selling personal protective equipment looking to set up operations in a new country.

Although SWOT is often criticised for being subjective and overly simplified, it can be a useful initial process in setting objectives for strategic planning and providing a basis for assessing core competencies and capabilities.

From this exercise, there would invariably be follow up work in devising suitable strategies and tactics to address the issues identified.

your business offering is certainly destined for failure! Important areas to consider may include:

- Demographics
- Market size and potential growth
- Customer wants and needs

- Motivation to buy the product
- Quantity and frequency of purchase
- Income level of customer.

Company

The company analysis entails an evaluation of business strategy, objectives and capabilities. At this stage, you should consider internal factors, including whether you have the business model for realising the delivery of your value proposition,

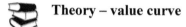 **Theory – value curve**

When comparing your organisation with the competition, it is important to undertake a comparison against key areas of value. In this respect, the concept of a value proposition can be useful – i.e. what are the key aspects of your product or service which offer value to potential clients, such as price or quality.

The value curve is an interesting qualitative approach to comparing the key areas of value between similar organisations. On the x axis list the key areas of value and on the y axis specific performance in relation to each value. The figure that follows shows a theoretical example of a value curve for two OSH training companies.

Although this approach has limited scientific grounding, it can be a helpful exercise in establishing the key attributes of your business that set you aside from the competition. Subsequently, attention can be given to key areas of value of perceived importance in developing market share.

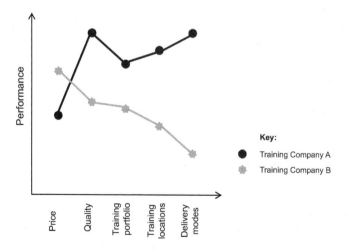

Figure 7.3 Example of a value curve

along with external factors, such as how well the company fits in with its external environment.

As a general guide, the following areas would typically be considered during the company analysis:

- Goals and objectives – what is the fundamental purpose of the business?
- Product line or service offering – what product/service are you offering to the market?
- Competencies – which capabilities do you hold with the greatest potential to create a competitive advantage?
- External considerations – how well does the organisation fit within the external environment?

Competitors

When preparing to go to market take time to evaluate who you are up against. A competitor analysis should take into consideration the position of competitors within the industry and the potential threat they may pose to your business. Analysis of both the current and potential nature and capabilities of a competitor will help you prepare to compete against them.

Existing firms in the market may take concrete steps to discourage new entrants from making moves to enter the market. Such steps, or strategic entry deterrence, can be any action towards creating or strengthening barriers to entry for the industry. These activities may be regarded as anti-competitive practices, opening a company up to scrutiny and violation of competition laws and regulations.

Many businesses have entered markets dominated by well established brands and found it very difficult to establish a foothold. Importantly, competitors will often react quickly to new entrants into a market and it is important to anticipate this response and be prepared.

Collaborators

Identifying and forging relationships with collaborators can help in building a competitive advantage. Collaborators can include:

- Agencies – many OSH consultancy and training businesses use agencies in various locations to help provide local knowledge and access to a client base
- Distributors – building a strong distribution network can create a barrier to entry for other companies
- Suppliers – as with distributors, suppliers are vital to the operations of a business. Strong relationships can be forged with suppliers using contracts and loyalties
- Partnerships – building an alliance with one or more strategic partners can be another way of building credibility in the workplace.

Arrangements for effective collaborations

If you are planning to team up with another organisation, be it a formal supply chain arrangement, or a more informal partnership, there are a number of important considerations that will aid effective collaboration:

- Perform due diligence – you will need assurance that the collaborator has the systems and capabilities needed before entering into any agreement
- Formalise agreements – even for more informal collaborations, it is good practice to formalise the understanding between each organisation. This could involve the application of a Memorandum of Understanding, detailing the purpose and objectives of the agreement along with roles and responsibilities
- Consider branding – when selecting a collaborator ensure the company is synonymous with your brand values and ethics
- Commitment – think deeply regarding the benefits for each party. Many collaborations struggle in the longer-term due to the commitment needed for an effective relationship, which clearly should be of value to both organisations.

Context

As discussed in Step 5 on *Economics and Financial Management*, the world of work is constantly changing and if you are providing a product or service to the market, it is important that you are aware of impending changes and the risks or opportunities they can create for your business. Consider the following questions:

- *Are there new technologies or trends that will impact me?* An awareness of new technology and its application to OSH can create substantial opportunities. For example, the growth of Information and Communication Technology (ICT) has introduced various opportunities for training providers, particularly linked to innovative modes of delivery, such as e-learning.
- *Are there any new laws or regulations that companies need to be aware of?* The constantly changing OSH legal landscape requires organisations to invest time and resources to keep up to speed with these developments. Technical indexes offered by organisations like IHS and Barbour are increasingly popular due to their ability to ensure that organisations keep up to date with changes in legislation and best practice.
- *Are there new risks and associated hazards being introduced into the workplace?* Changes in the world of work can often introduce new hazards and risks that require managing. Being proactive in creating solutions for managing emerging risks is another example of how you can quickly build a competitive advantage.

In relation to the context, it is important to identify opportunities in good time and act upon these quickly. The most successful organisations are often those with the capability to identify changes in the market before they occur.

Successful organisations are not afraid to invest resources to build capability – in this way they are able to steal a march on competitors and build market share.

Creating value

Businesses always like to talk about creating value. But what is value? Importantly, value is not just about price – although the price offered to a customer may be part of the value you are offering. Value, from a marketing perspective, is the difference between a prospective customer's evaluation of the benefits and costs of one product when compared with others (Macdivitt, 2011). In this sense, value can be expressed as a straightforward relationship between perceived benefits and perceived costs: Value = f [Perceived Benefits/Cost]. With respect to value, be aware of the key points of value that differentiate you from the competition and ensure that these are promoted to your customers.

Segmentation

Market segmentation is the process of defining and subdividing a large homogenous market into various segments. Importantly, market segmentation relates to consumers and not the product or service you are offering. A different marketing mix may then be employed for different segments. For example, many OSH training providers employ mass marketing techniques to promote their training courses. However, OSH training customers could be segmented into different areas (e.g. Government, Developer, Consultant, Contractor, etc.).

Targeting

Once market segmentation has been completed, a decision will be taken as to which markets will be targeted. As previously indicated, the market for OSH training could be made of various segments, with the decision as to which segment to target likely to consider the potential and commercial attractiveness of each segment.

Positioning

Positioning is developing your product or service brand image in the mind of the customer. What perception do you want the customer to hold regarding your product or service? This can in turn lead to the development of a value proposition. What are the key areas of value that you are delivering to potential customers? For example, if your company provides external OSH training, your key areas of value may include the quality of your training provision, or perhaps the range of training courses, or innovative delivery modes that you offer.

Once you have determined the key components of your value proposition, your business model needs to ensure that these areas of value are delivered to your clients (Anderson et al., 2006). After all, one of the most important characteristics of a successful business is delivering on your promises. If you market yourself as the leading provider of high quality OSH training but fail to deliver on this promise you will soon lose credibility in the market!

 When it comes to positioning it is often best to focus on a small number of core values and maintaining commitment to these choices.

Capturing value

Once you have put in place the strategy for creating value, the logical next phase is to make sure you implement measures to capture value. For this you can turn to the marketing mix of the "4 P's" – Product, Place, Promotion and Pricing (Milano, 2015). Importantly, the "4 P's" should not be considered in isolation. It is the interaction between these different choices which will often determine the success of your marketing strategy.

Product (or Service)

What exactly are you offering the customer and will it satisfy identified needs? A good discussion on the myriad of choices on product or service can be demonstrated with the provision of OSH training. For example, the following factors would need to be addressed:

- What level of training courses will be offered (foundation, intermediate, advanced, etc.)?
- What topics will the training courses address (industry specific hazards, organisational specific issues, etc.)?
- What delivery mode will be used for training provision (classroom, blended learning, distance learning, etc.)?
- What language will training be offered in?
- What accredited training courses will be offered (NEBOSH, IOSH, etc.)?

The above are just a few examples of common considerations, but remember always, it is not what *you* want to sell, it is what the *customer* wants to buy! You can have an excellent product or service, but you need to be sure there is a market for what you are selling.

Place

Where do you want to sell your product or service? Globalisation is resulting in many OSH organisations looking at expansion within new markets. For example, NEBOSH, the leading OSH awarding body, now attains most of its revenue from international qualifications. However, the decision as to where you look to expand your operations is often a complicated subject. First, you need to be sure that there is a market for what you are offering, but also that you have access to the right collaborators (suppliers, distributors, etc.) to ensure your choices are successful. What may have worked well in your home country may not always be successful in other parts of the world; so, take time to look at what the competition is doing and how you can learn and/or differentiate yourself from them.

Promotion

Promotion is the method you use to spread the word about your product or service to customers. The following are examples of how you could promote OSH products and/or services:

- Developing a website to showcase your product/services
- Creating a brochure detailing your offering
- Organising meetings with potential clients to introduce your product/service
- Publishing articles in recognised periodicals, ideally demonstrating an empirically proven need for you product/service
- Exhibiting at an OSH exhibition/conference.

When working on your promotional activities, it is important to consider where and when you can get your marketing message across to your target market. Also, think about when is the best time to promote. For example, timing promotions linked to recent developments in OSH legislation and best practice can help distinguish your offering from the competition.

 Take time to consider how your competitors (particularly successful ones!) promote their offerings; this may give you some insights on how to build promotional material.

Price

Unfortunately, there is no universal algorithm that you can apply to determine the pricing of your product or service. Pricing remains one of the most difficult aspects in marketing any product or service – this is no different for OSH. Pricing requires sound analysis and judgment and consideration of many of the factors addressed in the marketing analysis, such as competition and context.

Importantly, the relevant question for pricing is not asking simply "What price are our customers willing to pay?" Rather, the relevant questions are:

- "What is our product actually worth?" – our view
- "What value do our customers perceive?" – customer's view
- 'How can we bridge the gap between "reality" and perception?'
- And make a profit, given our costs!

The scenario presented above is an introduction to what is often referred to as value based marketing. As opposed to the typical cost based pricing model (incorporating a mark-up on costs), value based costing the price is determined by the value offered to a potential client, in line with a value proposition.

Pricing based on value all sounds very nice – but what can you do in situations where competition is largely price based? Try the following simple guidelines:

- Increase product differentiation and alert customers to risks of poor quality
- Change your price structure to improve value message
- Introduce "fighting" brands (a lower price offering designed to compete with low cost competition)
- Decrease comparability through complex price actions (e.g., bundling, non-linear pricing, loyalty programmes)
- Counter-attack in markets (geographic, products, customer segments) where competitors are vulnerable
- Decrease price through non-price variables (e.g., quantity discounts).

Sustaining value

The final part of your marketing strategy needs to consider measures for sustaining value. Indeed, the greatest organisations are not only able to achieve great business results but they are able to do this consistently year after year. At this stage, you should be focusing on customer acquisition and retention.

Acquisition

All businesses would like to increase their level of customer acquisition. Depending on specific needs, many organisations will look to implement a

Customer Relationship Management (CRM) system. This can be a useful tool in navigating through the sea of prospects to pinpoint and retain the right customers.

Although organisations spend inordinate amounts of money in trying to attract new customers, the return on investment is often greater when investing resources to retain customers. The key to acquisition is spending time with your customers and listening carefully to understand their needs. Remember, it is not what you want to be selling, but what the customer wants to buy! In many respects, this is going to require effective interpersonal skills to influence and persuade customers to select your product or service over the competition (see Step 10 on *Interpersonal Skills* for more discussion on soft skills).

Retention

Many companies search for an answer to retaining customers – the answer to this is simple – provide an excellent product or service! This may seem like a flippant comment, but far too often businesses spend too much time trying to chase down the next sale, without focusing on the commitments to deliver what they have in hand. If you really want people to come back to you, then you need to ensure that their experience is enjoyable.

However, it is not just a case of providing a good service – you need to ensure that customers understand that they have received a good service. Importantly, take time to communicate the value that the customer has received. Revisit your value proposition and remind customers of the benefits they have received in purchasing from your organisation.

Customer retention can also be enhanced by gathering information about the customer experience. This can be in the form of evaluation processes and possibly even self-assessment or reflection on the service provided. Clearly, it is not just a case of gathering this data, but making sure it is analysed and acted upon. It is important to remember that the best customers will not only continue to use your product or service but act as champions in promoting your brand to others. A useful mantra in trying to get to this position is to "under promise and over deliver."

Internal implications for OSH marketing

Most of the discussion presented in this Step is focused on the marketing of OSH products and services to external customers. However, many of the lessons from this Step can be applied if your OSH role does not involve a commercial dimension. In fact, if internal stakeholders see effectiveness in the delivery of OSH services and the value that you provide then you are more likely to have their continued support.

At the start of the Step, the comment was made that OSH practitioners need to become better at getting people interested in OSH as it is not always regarded as an exciting subject. In some countries OSH is perceived as something that gets in the way of industry; therefore, it is incumbent on OSH practitioners to work hard to change this perception and improve the credibility of the OSH profession. We must be creative in the ways in which we sell OSH. Importantly, the term "sell" is used intentionally, as we are all marketeers looking to get people to buy into OSH.

Launching an OSH campaign

One example of when you will have the chance to sell OSH in your organisation and to wider stakeholders is the launch of an OSH strategy. Previous discussion on OSH strategy in Step 1 on *Strategy and Leadership* indicated that an effective OSH strategy is typically accompanied by a compelling OSH vision and there may be opportunities to deploy your strategy through an OSH campaign. Indeed, many of the organisations that embrace zero based OSH visions have introduced powerful campaigns to help bring these visions to life.

Importantly, the launch of OSH campaigns will often take place within the context of existing brand architecture. Marketing departments will typically be protective of your organisation's existing brand and may be reluctant to introduce new brands which may cause confusion with customers. However, by working closely with your marketing department you should be able to create a visual identity that can leverage the existing brand architecture.

The launch of an OSH strategy through a campaign will entail the development of guidelines on how the visual identity and messaging will be used in the organisation. This can include guidance on how any logos, icons or taglines should look on personal protective equipment, clothing, stationary and other items, such as

---

 Effective OSH campaigns

Abu Dhabi Occupational Safety and Health Center (OSHAD), the regulator for health and safety standards in the Emirate of Abu Dhabi, implements various campaigns and other innovative approaches to providing OSH information to workers whose first language is not English, or may have low literacy levels.

The Safety in Heat (OSHAD, 2016) programme is a great example of one of these campaigns, providing a range of materials, such as posters and leaflets for employers, managers and employees to assist with the management of risks associated with working in the heat. The information is free of charge and provided in a variety of languages, including Arabic, Hindi, Urdu, Tagalog and Bengali.

A key requirement for an effective campaign and associated communication is ensuring that it is understandable to all relevant workers and interested parties, so take time to ensure that your OSH messages are being communicated in an appropriate way.

marking on vehicles. The application of marketing principles to ensure that your OSH strategy is launched in a consistent and professional manner can have a powerful impact on how it will be received in your organisation.

Developing your personal brand

Not only do you need to market OSH within your organisation; you also need to ensure that your role as an OSH professional is communicated. This is about developing your personal brand and being able to market yourself effectively (Allen, 2014). In some respects, the marketing of your role is more important than other roles in an organisation, as it may be perceived as a position which does not provide financial value to the organisation.

> ✅ *Spend 80% of the time working on OSH issues and 20% of your time communicating the benefits of these activities to the organisation.*

In step 10 on *Interpersonal Skills* we explore the attributes needed to influence and persuade others. Communication skills are intrinsically linked to being a good marketeer, so make sure that you are communicating the benefits of your work. As a rule, try to carve out time in your working day to articulate the

 Elevator pitch

The idea behind the elevator pitch is that you find yourself in an elevator with a person who you have not met before and you have the time it takes for the elevator to reach the ground floor to articulate your personal brand.

Your personal pitch is a succinct statement that clearly communicates what you do and how others can benefit. It gives enough information so that it is clear what you are talking about and it makes others want to know more.

Spend some time reflecting on your personal pitch in consideration of the following characteristics:

- Present, past and wfuture
- Targeted to audience
- Succinct, easy to understand, credible
- Communicated with passion and energy
- Length – 10 to 60 seconds – 200 words max

value of your activities and do this in a way that that is interesting, innovative and appeals to those it is intended for. This will ensure that people are regularly informed of what you and the OSH function are doing and why it is important for your organisation.

Summary

Marketing is the process of selling something, to somebody, somewhere; and without an effective marketing strategy that can be executed in practice your organisation is unlikely to meet its goals. It is important to recognise that marketing is a very broad subject encompassing various strategic and tactical decisions that are intended to create value.

If you are selling OSH products or services then it is necessary to take time to analyse your market, the sectors that you wish to operate in and the clients that you wish to do business with. If you can offer something that is unique or different, position your offering effectively to the market and have the resources and capabilities to deliver on your promises then you have the foundations of a successful business.

For those of you who are not selling products or services, understanding marketing principles and concepts is still extremely valuable. Some of the most influential OSH practitioners are evangelists. This may seem like a strange term but the ability to "sell" OSH both within your organisation and externally to different stakeholders that you interact with can have a significant impact on personal and organisational success. So, make sure you are spending sufficient time communicating the value of OSH in your organisation.

Key takeaways

- Marketing is a strategic activity concerned with identifying needs and providing value for customers, partners and society – *have you established a marketing strategy for the OSH products and/or services that you provide?*
- Marketing analysis involves a study of the various internal and external factors to gauge current position and capability with respect to the intended product and/or service offering – *does your marketing analysis sufficiently address the "5 C's"?*
- The creation of a value proposition, market segmentation, targeting, positioning and capturing value are essential aspects of marketing – *have you adopted a structured and systematic approach to your OSH marketing strategy?*
- Marketing OSH internally is important to raise awareness and interest – *are you spending enough time communicating the value of OSH in your organisation and do you have a compelling campaign for your OSH strategy?*
- As an OSH practitioner, the creation of a personal pitch is an important activity in communicating your personal value – *have you taken time to reflect and define your personal pitch?*

References

Abu Dhabi Occupational Safety and Health Center (OSHAD). (n.d.). *Safety in Heat.* Retrieved from https://www.oshad.ae/safetyinheat/en/index.php.

Allen, R. B. (2014). *Personal branding and marketing yourself: The three PS marketing technique as a guide to career empowerment.* Waltham, MA: Balian Publishing Co.

Anderson, J. C., Narus, J. A., & Roussum, W. V. (2006). Customer value propositions in business markets. *Harvard Business Review.*

Chernev, A. (2014). *The marketing plan handbook.* Chicago, IL: Cerebellum Press, 4th edition.

Dolan, R. J. (2000). *Notes on marketing strategy.* Harvard Business School Background Note 598–061.

Kumar, N. (2004). *Marketing as strategy: Understanding the CEO's agenda for driving growth and innovation.* Cambridge, MA: Harvard Business Press, 1st edition.

Macdivitt, H. (2011). *Value-based pricing: Drive sales and boost your bottom line by creating, communicating and capturing customer value.* Columbus, OH: McGraw-Hill Education, 1st edition.

Milano, C. (2015). The marketing mix: Master the 4 Ps of marketing. *50Minutes.com.*

Seth, C. (2015). The SWOT analysis: Develop strengths to decrease the weaknesses of your business. *50Minutes.com.*

Part III

Distinctive

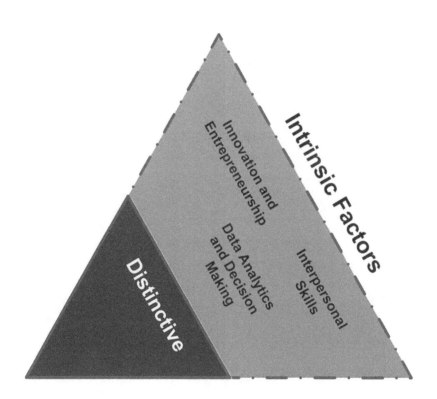

8 Data analytics and decision making

Step 8 is the first Step that relates to our third and final theme in the OSH Practitioner Transformation Model on the need to "be distinctive". In this Step, we cover data analytics and decision making, which may appear somewhat unusual content for the theme of being distinctive; however, the effective use of statistics can help bring data to life and ensure that OSH messages are communicated in a compelling fashion. Competency in data analytics will also help to bring greater objectivity and credibility to the OSH information that you present to others and the decisions that you make.

On completion of Step 8, you will be able to:

- Be comfortable with uncertainty and know how to deal with it
- Understand analytical descriptive, summary and inferential statistical tools and their purpose and limitations
- Identify different ways to present OSH data in consideration of the target audience
- Explain how to incorporate OSH data into management reports
- Recognise the contribution of luck in OSH outcomes and whether it is possible to create an environment that increases the propensity for luck.

Managing uncertainty

Business decisions are often too complex to be made by intuition alone. Understanding the risk involved with the support of data and other quantitative information can be crucial to clarify the available options when making decisions. Quantitative information is everywhere in business: costs, revenue levels, productivity figures, customer satisfaction ratings, the list goes on. Within the world of big data, we are often faced with extremely large data sets creating challenges to make sense of a mass of quantitative information.

As an OSH practitioner, it is extremely important that you are comfortable with uncertainty and capable of using statistics and analytical tools to enable better informed decision making. Indeed, every day in life we are faced with uncertainty when making decisions, often based on limited information. This is no different in the profession and discipline of OSH, where decisions can often be a matter of life or death. This raises an intriguing question as to whether you feel that you are a good decision maker? So much of the decision making conducted by OSH practitioners is risk based; but do you feel confident in the logic or intuition that you combine in making decisions? Invariably, good decision-making is often a combination of gut feeling and objective data. However, do not rely on intuition alone, as it is often a poor guide when making decisions.

 Good decision making requires "gut instinct" supported by credible, objective data to help justify your decision.

Data analytics and OSH

So why do we need data and data analytics? Fundamentally, it is important in life and in business that we make informed decisions. Being more informed and aware is critically important in situations where our decisions impact OSH outcomes. OSH practitioners regularly use data to present OSH performance. Subsequently, it is important that OSH practitioners can assess the validity of data to present a credible picture of OSH performance. Data may also be compared to data from similar organisations for benchmarking purposes to learn lessons and identify improvement areas, again emphasising the importance of valid, reliable and credible information.

 79% of respondents either "agreed", or "strongly agreed" that I have a good understanding of statistical methods and their application for the analysis of OSH data.

Statistics

There are two broad categories of statistics: namely descriptive and inferential statistics (Loether & McTavish, 1998). Descriptive statistics, as the name suggests, describe or summarise data in a meaningful way to help identify patterns that may arise from data. Inferential statistics enable conclusions to be drawn about a population when only a sample of the population is considered. In many cases, it is not feasible to access an entire population and inference will be drawn on limited information; it is therefore important that the sample closely reflects the population.

Descriptive statistics

When the types of OSH incidents vary, the use of descriptive statistics can help in presenting data and identifying trends. Descriptive statistics contain information on *frequencies*, *averages* and *standard deviation* which are explained in the following sections. Data is normally presented in a table, referred to as a frequency table, identifying the frequency of incidence, as demonstrated in the simple example showing different types of OSH incidents in Table 8.1.

In addition to frequency tables, data can be presented diagrammatically using bar charts, pie charts and line graphs. Where data is grouped, as below, the representation of frequency information is known as a histogram.

Table 8.1 Frequency table of Occupational Safety and Health incidents

2011	Jan	Feb	Mar
Machinery contact	2	0	1
Struck by	0	0	0
Struck against	2	2	2
Slip or trip	2	1	2
Fall	0	0	0
Manual handling	4	3	3
Electricity	0	0	0
Harmful substances	1	0	0
Fire/explosion	0	0	0
Other	0	0	1

Source: Original – developed by Rob Cooling

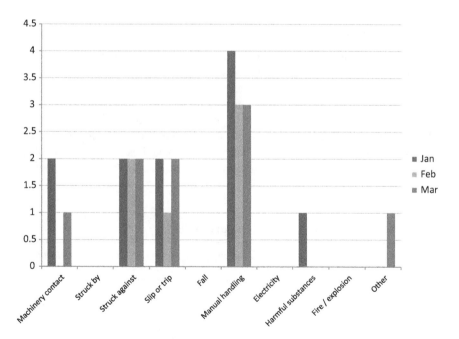

Figure 8.1 Histogram of Occupational Safety and Health incidents

Source: Original – developed by Rob Cooling

Measures of central tendency

When looking at a set of OSH data there will often be interest in measures of central tendency. The central tendency is the term used to indicate the mid-point of a set of numbers along with the mean, median and mode. These can be demonstrated using the data in the table below.

The mean is commonly termed the average. From the above table, the mean days lost is the total number of days lost divided by the number of accidents causing lost time. 35 / 17 = 2.06 therefore the mean number of days off is 2.06. The median is the middle value in a data series. Where there is an uneven number of data series then the median is simply the mid-data, however, where there is an even number of data series then the mean of the two middle data values is taken. From the above table, the "middle" of the 17 accidents is between the 8th and 9th, which gives a median of 1.5-day lost time. The mode is the data with the largest frequency, so from Table 8.2 the most frequent number of accidents is 7, therefore the mode of lost days is 1.

The use of mean, median or mode for the central tendency will depend upon the manner of presentation. In the previous example, it would be unlikely that the mean value of 2.06 days lost would be used, as whole or half day representation is more common. The mean value is also the most sensitive measure as it is affected more by extremes of data. The median may be more appropriate where fractional values are not realistic, whereas the mode could be used to represent a value that is more likely to be repeated in the future.

The flaw of averages

Measures of central tendency can help in providing useful summary data, however, unfortunately, the flaw of averages is that plans based on assumptions about average conditions usually go wrong! Take for example, evaluating summary data on average fines for OSH offences within a particular jurisdiction. Although there could be numerous OSH fines around the mean, median or mode, it would only take one extremely large fine to skew this data and impact the validity of a measure of central tendency. Therefore, when analysing data, it is important that OSH practitioners are tuned into a range of different values (a distribution) as opposed to single values.

Table 8.2 Data on Occupational Safety and Health incidents

Number of days lost per accident (x)	Number of accidents causing lost time (f)	Person days lost (fx)
1	7	7
2	4	8
3	2	6
4	1	4
5	2	10
Total 15	Total 16	Total 35

Range

The range is simply the difference between the largest and smallest values in a data set. For example, using the data set below the range would be calculated as follows:

Table 8.3 Additional data on Occupational Safety and Health incidents

Type of incident	Electricity	Hazardous substances	Fall	Machinery contact	Moving vehicle	Fire	Manual handling	Slip/trip
No. of incidents	2	4	4	4	5	5	7	9

Using the data above the range would be calculated as $9 - 2 = 7$

Whilst using the range as a measure of spread is limited, it does set the boundaries of the scores. This can be useful if you are measuring a variable that has either a critical low or high threshold (or both) that should not be crossed. The range will quickly provide information whether at least one value breaks these critical thresholds.

Standard deviation

Standard deviation is a number used to tell how measurements for a group are spread out from the average (mean), or expected value. A low standard deviation means that most of the numbers are very close to the average and a high standard deviation means the numbers are spread out.

Consider the data shown in Table 8.3 on the total OSH incidents at an organisation during a year.

These 8 numbers have the average (mean) of 5

$$\frac{2 + 4 + 4 + 4 + 5 + 5 + 7 + 9}{8} = 5$$

To calculate the population standard deviation, it is required to find the difference of each number from the mean then square the result of each difference:

Table 8.4 Standard deviation of Occupational Safety and Health incidents

Value – Mean	Total	Total squared
2 – 5	−3	9
4 – 5	−1	1
4 – 5	−1	1
4 – 5	−1	1
5 – 5	0	0
5 – 5	0	0
7 – 5	2	4
9 – 5	4	16

Next, find the average of these values and then take the square root.

$$\sqrt{\frac{9 + 1 + 1 + 1 + 0 + 0 + 4 + 16}{8}} = 2$$

The above answer is the population standard deviation. The formula is only true if the eight numbers that we started with are the whole group. If they are only part of the group picked at random, then we should use 7 (n − 1) instead of 8 (which is n) in the bottom (denominator) of the second last step. This would be referred to as the sample standard deviation. Whereas the standard deviation of the sample is the degree to which individuals within the sample differ from the sample mean, the estimate of how far the sample mean is likely to be from the population mean is referred to as the standard error of the sample mean.

Increasingly, companies are turning to software-based cures for the flaw of averages. Many programs now simulate uncertainty, generating thousands of possible values for a given scenario – in essence, replacing the snapshot of a single average number with a more balanced picture comprising a whole range of possible values and their likelihood of occurring – the frequency distribution. A popular tool is Monte Carlo simulation, with spreadsheet-based simulation software widely available and used in fields as diverse as petroleum exploration, financial engineering, defence, banking and retirement portfolio planning (Savage, 2002).

Sensitivity analysis

Sensitivity analysis is a technique used to determine how different values of an independent variable impact a dependent variable under a given set of assumptions (Saltelli et al., 2009). Sensitivity analysis is commonly used when assessing the results from a cost-benefit analysis (see Step 5 on *Economics and Financial Management* for more information on cost-benefit analysis). For example, when assessing the benefits of proposed OSH improvement programs, the potential results will be subject to uncertainty.

Regression analysis and forecasting

Linear regression models are used to determine the relationships between variables that are believed to be related. Once a relationship is established it can then be used to forecast the future (Wilson, 2012). This could be particularly useful when analysing OSH data and attempting to draw inference regarding future incident trends. Regression analysis can also be used to answer questions such as "What is the effect of working hours on the number of workplace incidents?". The independent variable (X) in this scenario is the number of working hours, i.e. the variable which is believed to cause the other variable to happen. The other variable is the dependent variable (Y), which in this case is workplace incidents. The number of hours affects the number of workplace incidents and not vice-versa.

Regression analysis involves gathering sufficient data to determine the relationship between the variables. With many data points, such as years' worth of information on working hours and numbers of workplace incidents, a graph can be drawn with the number of working hours on the X axis and the number of workplace incidents on the Y axis. The goal of regression is to produce an equation of a line that best depicts this relationship. Regression tries to fit a line between the plotted data points so that the squared differences between the points and the line are the least.

R Square

The R Square is a measure of how close the data is to the regression line. R Square is always between 0 and 100%:

- 0% indicates that the model explains none of the variability of the response data around its mean
- 100% indicates that the model explains all the variability of the response data around its mean
- In general, the higher the R Square, the better the model fits your data.

R Square provides useful information of how well a linear model fits a set of observations. However, R Square needs to be supported by other measures to establish a formal hypothesis test for any relationship.

T Test

The T Test can help determine if the regression equation is effective to use for forecasting. More specifically, the T Test reveals if an X variable has a statistically significant effect on the Y variable, such as our previous example looking at the effect of the number of working hours on the number of workplace incidents. When considering whether a model is valid for forecasting purposes it is necessary to have both a high R Square and a high T statistic (Silbiger, 2005).

Presenting data

The way we present data is important and depends on various factors and most importantly the intended audience. Graphical representation of data is often a favoured method as it is quick at explaining varied issues; as the saying goes "a picture speaks a thousand words". The use of diagrams, flow-charts, graphs, pictograms, infographics and sometimes even photographs can be extremely powerful, but should be used wisely to avoid over-communication with graphical representation. A balance of visual representation supported by effective narrative explanations is often preferred, with the application of colours and diagram keys carefully embedded.

The following table outlines a range of different types of visual instrument and the associated strengths, opportunities and limitations with these different methods.

Table 8.5 Evaluation of mechanisms for presenting Occupational Safety and Health data

Type of visual instrument	Key strengths	Opportunities	Limitations
Tables	Useful for presenting data with specific amounts	Effective in dealing with many different units of measurement Easier to extract specific details	Takes longer to read and comprehend due to structure Difficult to capture trends
Line graphs	Show changes over time	Can compare multiple continuous data sets easily Interim data can be inferred	Only to be used with continuous data
Bar charts	Useful for absolute values and contrasts between areas and places	Good visual representation of statistical data Simple to construct and easy to understand	Graph categories need to be reordered to emphasise certain effects Suitable only for discrete data Limited space for labelling with vertical bar graphs
Scatter graphs	Used to compare two sets of data	Effective in showing a correlation between two data sets Relatively east to construct Show data spread clearly and any anomalies stand out	Too few data points can produce skewed results Difficult to label points and ascertain exact values Too many data points can impact readability Cannot show relationship between more than two variables at once
Pie Charts	Useful for %s and showing statistical data	Shows % of each segment Easy to create Can be used to represent a wide range of statistical data and are visually effective	Many segments can make the charts cumbersome Calculation of the amounts can be difficult Many small segments can be difficult to analyse

Interpreting data

The limitations of interpreting results are twofold: over generalisation of the results and misinterpretation. As statistical data is historic it does not necessarily indicate future performance. As previously discussed, the data from any study only applies for the population studied under the conditions at the time of the study. Therefore, competence in the theories of causation and data dimension variables is needed to avoid any generalisation beyond these parameters.

A study only identifies patterns of data distribution and assumptions must not be made about the correlation between data without considering the variables for

Hans Rosling

On 7 September 2017, Hans Rosling, the Swedish physician, academic, statistician and public speaker, sadly passed away. Rosling, a professor of global health at Sweden's Karolinska Institute, produced work dispelling common myths about the so-called developing world, showing that most of the Third World is on the same trajectory toward health and prosperity, and many countries are moving twice as fast as the west.

Not only was Rosling a global health expert, he was also a data visionary. What set Rosling apart weren't just his observations of broad social and economic trends, but the stunning way he presented them. Presentations that track global health and poverty trends are often boring, but Rosling brought big data and the big picture to life.

OSH information may not always be the most interesting, but it is in your hands to look at creative and innovative ways of presenting this information so people sit up and take notice.

So, why not spend some time watching Hans Rosling's famous TED talk (TED, 2006) to help you to come up with news ways of presenting OSH data?

a plausible causal link; this may require further trial and error investigation of existing or further data. For example, if a study of incidents found that there was correlation between those deemed as sociable and outgoing (extroverts) and the numbers of recorded injuries, care should be taken before drawing conclusions from this data. In this case, it may be tempting to assume that extroverts have more accidents, however, further studies would be required, which might demonstrate that extroverts are likely to report injuries, demonstrating a link between personality traits and the reporting of injuries rather than their occurrence.

 Remember that commonality is not causality; further studies will invariably be required to establish a causal link between data.

When analysing trends, it should be determined whether any change of direction of the trend is more than just a chance or random fluctuation. For example, if an organisation manufactured 500 components a day and in each year had 100 accidents recorded, then the following year, given that nothing had changed with the manufacturing process, you could expect a figure of around 100 accidents per year. If the case arose whereby 101 or 99 accidents were recorded in a year then this could be down to random fluctuation. However, if figures were recorded of 80 or 120 accidents a year then the difference would be more significant from the expected figure of 100. For long term data analysis, identifying upper and lower limits of significance of deviation, the trend can be calculated and plotted. Deviation of the trend beyond these significant boundaries would indicate external influences beyond that of random fluctuation.

One use of accident and incident rates is to enable trend analysis to be completed when the organisation is facing a number of different changes since the parameters within the organisation for calculating the rates remain the same and an indication of the movements of those rates beyond significant boundaries can be determined. When using accident and incident rates as a measure or comparator of an organisation's OSH performance there must be clearly drawn guidelines as to definitions, interpretation and multipliers that are used. The level of risk between the organisations must be comparable to successfully use these rates for comparisons.

With respect to benchmarking of OSH data, typically comparisons are made between parts of the same organisation, between similar organisations and between organisations within a relevant industry or service sector, often across different countries. As indicated above, all these comparisons would be subject to clearly defined terminology and boundaries in the scope of the comparison. It is also worth noting that benchmarking of OSH standards, culture and performance often takes place qualitatively, whereby organisations compare different approaches to OSH management and associated outcomes.

 Evidence-based management

Every day we implement OSH measures which we believe will have impact in improving OSH performance. Inordinate amounts of time and money are spent on leadership initiatives, communication campaigns and training courses; but are we confident that we are spending money in the right areas? One approach that has been advocated as a way of improving decision making is evidence-based management (Pfeffer & Sutton, 2006).

Evidence-based management, a concept which originated in the medical profession, centres on the understanding that decisions should be based on the latest and best knowledge of what works. This may seem like a nonsensical statement, as what else should guide our decisions? However, the reality is that many OSH decisions are linked to long-standing practices, patterns gleaned from experience and methods which have been implemented in other organisations.

Ideally, evidence-based practice should be established from research conducted in your own organisation. In addition, every year, thousands of research studies are conducted in the field of OSH which help to provide and improve critical thinking and decision making.

Management reports

OSH practitioners will often have to incorporate data into reports for management attention. It is important that these reports are structured appropriately, with consideration of the following:

- Executive summary – Describing the purpose, key findings and recommendations of the study. Presented in a high level and business focused manner for busy Executives who may not have time to read the full report
- Introduction – Providing background on the study along with aims, objectives and the scope of the study
- Data collection – Defining how the data was collected and explaining whether the data is primary (i.e. collected by yourself) or secondary (i.e. collected by someone else). This section will describe if data is collected through survey, or other mechanisms, and how this data is converted into either quantitative and/or qualitative outputs
- Data analysis – Detailing whether quantitative (i.e. descriptive and inferential statistics are used) and/or qualitative techniques are used to analyse the data
- Discussion – Analysing the data collected in the context of the problem statement or issue that is being addressed. This is a critical stage of the study as it requires the simplification and interpretation of data
- Conclusions and recommendations – Summarising findings and providing recommendations to address the initial issue or problem established. Recommendations are supported by action plans, details of actions to be undertaken, responsibilities, resources required and timescales for implementation of actions

OSH and luck

Until now the focus in this Step has been on trying to make sense of the world based on information and data that we are presented. However, in the world of OSH not everything is under our control and sometimes luck can be a factor in the occurrence of loss events, just as it is in the realisation of opportunities. Luck (or serendipity) is an interesting word in OSH, relating to a chance encounter or incident that leads to a satisfactory (or unsatisfactory) and sometimes business transforming conclusion. Although there is no doubt that luck is attributed as a common factor in many OSH incidents, there is presently no discernible model which considers how luck may be created and acted upon in business.

To develop a framework or model for managing luck, it would be necessary to select a large number of organisations for whom the consensus opinion is that luck played a significant factor in incident causation. Subsequent analysis of these incidents may help in determining underlying characteristics which may contribute to an environment for fostering luck (Rovira, 2004). For example, there may be common characteristics across organisations, such as conscientiousness, collaboration, openness, etc., which could increase the potential for luck.

In some cases, it is apparent that organisations do not recognise that they are experiencing a serendipituous event until it is too late. Furthermore, recognising and responding to chance events may not be an internal competency present in all organisations. It would therefore be necessary to implement specific arrangements to build an organisational culture with the flexibility and responsiveness to capture the benefits of chance events.

Luck is anything but an exact science, however, the correlation between specific organisational characteristics and the ability to capitalise on chance events is an interesting discussion point. I am sure that there will be times when luck has played an important role in your personal and professional life and it may be worthwhile to give more consideration in terms of how it impacts OSH outcomes.

Summary

To become a better-informed decision-maker it is essential that you are comfortable with statistics and analytical tools. Ultimately, this will mean that you also become more comfortable with uncertainty, which is at the heart of the discipline and profession of OSH. Unfortunately, despite the effectiveness of your OSH strategies and programs there will always be uncertainty; but learning how to quantify and present uncertainty and knowing the right questions to ask when presented with OSH data will help ensure that you make decisions which are well-informed. They may not always be the right decisions but at least you will be able to show that you have considered all the relevant and available information objectively and in a robust manner.

So, start trying to learn a bit more about the various ways to perform data analysis today. It may not be the "sexiest" aspect of OSH, if indeed OSH does have any sexy elements, however, competency in this area will help you become more *Distinctive* as an OSH practitioner in presenting a strong justification for your OSH decisions, and as Benjamin Graham famously said: 'You are neither right nor wrong because the crowd disagrees with you. You are right because your data and reasoning are right'.

Key takeaways

- OSH risk management is essentially uncertainty management as we are dealing with situations related to the prevention of future outcomes based on current information – *are you comfortable with uncertainty and how it can be quantified and presented?*
- OSH quantitative data can be analysed using a range of descriptive, summary and inferential statistics – *can you use a range of statistical techniques to analyse OSH data?*
- The presentation of OSH data is important to ensure that information is read and absorbed – *are you using a range of approaches to present OSH data in consideration of the target audience?*

Measuring OSH culture

The strength of an organisation's OSH culture is largely recognised as a key determinant of OSH performance. Subsequently, it is important to spend time measuring and analysing your OSH culture. There are a number of proprietary models available for measuring OSH culture, however, the elements of an effective OSH culture are widely recognised and publicised, such as leadership, employee involvement, communication and competency.

When considering these elements, think about the specific dimensions that you could measure. For example, communication could be measured in different ways, such as the metrics related to communication activities, the reach of communication, the level of engagement or the impact from communication.

A thoughtful approach to determining metrics in line with your OSH strategies and goals will ensure that your organisation really measures what matters.

- The reporting on OSH data should be conducted in a structured and systematic manner – *do your management reports incorporate sufficient data analysis and present this information in a compelling fashion?*
- Luck is a common factor in OSH incidents and the prevalence of serendipity in OSH should not be understated – *have you considered how an environment could be created to encourage luck?*

References

Loether, H., & McTavish, D. (1998). *Descriptive and inferential statistics*. Saddle River, NJ: Prentice Hall, 3rd edition.

Pfeffer, J., & Sutton, R. (2006). Evidence-based management. *Harvard Business Review*. Retrieved from https://hbr.org/2006/01/evidence-based-management.

Rovira, A. (2004). *Good luck: Creating the conditions for success in life and business*. San Francisco, CA: Jossey-Bass.

Saltelli, A., Chan, K., & Scott, E. M. (2009). *Sensitivity analysis*. Hoboken, NJ: Wiley, 1st edition.

Savage, S. (2002). The flaw of averages. *Harvard Business Review*. Retrieved from https://hbr.org/2002/11/the-flaw-of-averages.

Silbiger, S. (2005). *The 10-day MBA: A step-by-step guide to mastering the skills taught in top business schools*. New York, NY: William Morrow & Co.

TED. Ideas worth spreading. (2006). Retrieved from www.ted.com/talks/hans_rosling_shows_the_best_stats_you_ve_ever_seen.

Wilson, J. H. (2012). *Regression analysis: Understanding and building business and economic models using excel*. New York, NY: Business Expert Press.

9 Innovation and entrepreneurship

In Step 9 the application of innovation and entrepreneurship to OSH is considered. Discussion addresses the concept of innovation and how an OSH innovation typology can be established to help in discovering new ways of doing things differently. Consideration is also given to change management and the approach needed to ensure that your new ideas generate buy-in and gain momentum. The growth of Information and Communication Technologies (ICT) will also be addressed, with evaluation of the threats and opportunities created by the rapidly moving digital world. With the increasing number of OSH practitioners looking to escape from the corporate world to establish their own businesses, the Step will also provide learning points on how to embrace entrepreneurial spirit and develop a compelling proposition for your own business.

On completion of Step 9, you will be able to:

- Explain what is meant by the term innovation and the concept of an innovation typology and how it can be applied to OSH
- Identify the importance of adopting a strategic approach to OSH change management
- Discuss the growth of ICT and its implications for OSH management and practice
- Recognise the different types of social media and the benefits and limitations for OSH practitioners
- Understand what's involved in starting your own OSH business.

If there was one thing that as an OSH practitioner you would want to be known for in your organisation, surely being innovative would be high up on the list. However, far too often a copy and paste approach is adopted towards OSH management; systems, processes and tools that have been used in the past are regurgitated and relabelled.

This does not imply that we should always look to reinvent the wheel, as existing frameworks and methodologies can provide an excellent starting point in establishing OSH management arrangements. But it is important not to become a slave to familiar approaches. Distinctive OSH practitioners embrace innovation and challenge the status quo in looking for new and improved ways to achieve OSH goals.

Innovation is the practice of making changes, particularly by introducing new methods, ideas, or services. When you start thinking about innovation, companies such as Google, Facebook and Twitter may come to mind; but innovation is not only applicable in highly technical and fast-moving sectors. It is valuable in all businesses and associated functions to possess unique resources and capabilities.

Opportunities for OSH innovation

 89% of respondents either "agreed", or "strongly agreed" that innovation is very important to be successful as an OSH practitioner.

Our OSH Practitioner Insight Survey identified that 89% of respondents either "strongly agreed" or "agreed" that innovation is very important to be successful as an OSH practitioner. The ability to identify new ways of doings things is fundamental to continual improvement in OSH metrics. Many organisations find that their OSH performance can plateau and they often must adopt new approaches to generate further improvements. So, take time to think about what you are doing to differentiate your approach to OSH practice from others.

Creating on OSH innovation typology

A good starting point in uncovering opportunities for innovation is to develop an OSH innovation typology. A typology is a way of putting things into groups (or types) according to how similar they are and it can be a valuable way of considering the interrelated dimensions of OSH practice. Figure 9.1 shows an example of an OSH innovation typology. It attempts to categorise forms of OSH innovation which have the potential to improve performance.

The different types of innovation are discussed in the following sections in the context of OSH – why not try applying this approach in your organisation?

As presented in the model above, there are a range of different types of innovation which are applicable to OSH:

- Framework/methodology – Innovation can be introduced in the frameworks and methodologies used for OSH strategy and management. This may not always entail an extensive overhaul of existing approaches, but tweaking and fine-tuning to bring about improvements in effectiveness and efficiency. Some

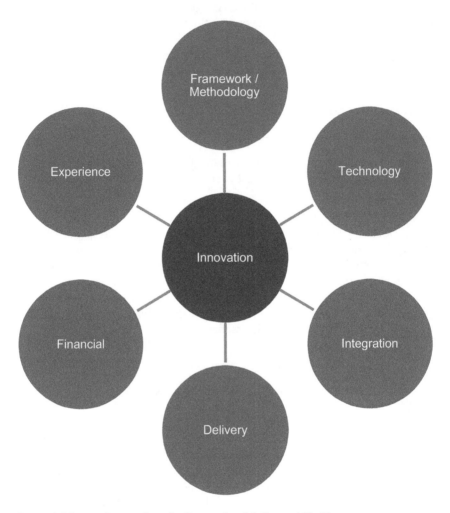

Figure 9.1 Innovation typology for Occupational Safety and Health

organisations have developed strong reputations for the innovative methods they use for OSH practice and have even commercialised these approaches for provision to external clients.

- Technology – With the continuing growth of information and communication technologies, including social networking, innovation through technology represents another way to advance OSH performance. This includes the creation of applications for smartphones or tablets to conduct site inspections, surveys or audits. Some organisations may also be able to leverage IT departments' expertise by developing online tools and other solutions.

- Integration – Enhancing integration of OSH with other business disciplines and functions is another common form of innovation. Many organisations operate with a functional structure which fosters cohesion within the boundaries of the function, but can lead to silos and over-attention to function-specific issues. Innovation through integration explores ways to encourage greater collaboration, such as closer working between OSH and related business functions, such as human resources, or even the introduction of matrix structures.
- Delivery – Innovation can arise across the various dimensions of delivering OSH support services in an organisation. For example, in communications, there are various innovative ways of delivering OSH messages, such as the use of industrial theatre or the provision of visual impact training. In OSH training, a variety of delivery mechanisms can help develop levels of OSH awareness and competency, such as online portals and blended learning.
- Financial – Innovation in finance and OSH include the presentation of advanced business cases for OSH expenditure, such as return on investment and payback analysis which helps determine the break-even point of an investment. Innovative practice in this area may increase buy-in and interest in OSH from top management.
- Experience – Innovation through changing employee experience is strongly linked to OSH culture. Do people experience OSH as associated with red-tape and bureaucracy, or is it less about documenting OSH and more about meaningful conversations with people to understand priorities and barriers? OSH cultural change programmes and associated interventions can be an effective way of improving employees' experience.

Reflective practice is an important element in developing an OSH typology as insufficient time is often taken for reflection and engagement to understand opportunities for innovation. The development of knowledge management systems can also play an important role in establishing your OSH typology, by ensuring key lessons are recorded and communicated (Budworth & Ghanem, 2014).

 Reflective practice is important in developing an OSH typology so try to find time for reflection and engagement to understand opportunities for engagement.

A dichotomy?

For innovation to prosper, a culture needs to be developed in which employees are encouraged to "think outside the box". However, this creative thinking may be hard to square with the order and consistency needed for effective OSH

Industrial Theatre

In some parts of the world, such as the Middle East, there are significant challenges in conveying OSH messages in the construction industry to a workforce largely comprised of migrant labour, who may have poor levels of literacy.

To ensure the importance of OSH is understood by the workforce there is often a need to look at innovative approaches to learning and development. One particular example is the concept of Industrial Theatre entailing the use of theatrical performances of OSH scenarios to highlight the importance of OSH on the job site and convey how certain behaviours can lead to risk, and how, in turn, risk can be avoided.

Industrial Theatre often involves actual workers to present experiences of what happens on site with elements of humour to provide a relatable performance for the audience. In addition, props and other learning aids are used to replicate actual site activities and conditions.

Industrial Theatre is a great example showing that innovation is not just about technology and that there may be many opportunities to innovate across the development, implementation and maintenance of OSH management systems and arrangements.

management, particularly in high reliability organisations, which must manage complex hazards. Furthermore, management by objectives and organisational culture can create inertia that may stifle innovation. For example, if the organisational focus is on what is needed now, it may be difficult to create a mind-set to inquire what is needed in the future (Christensen, 2011).

Creating an environment to encourage innovation is challenging. Some organisations have even been known to set aside time during the working day to encourage employees to focus solely on innovation. For example, Google famously permitted its engineers to spend 20% of their time on personal projects to empower them to be more creative and innovative (Gersch, 2013). Allocating even small blocks of time in your calendar may help in encouraging reflective practice, but the wider issue is how to develop a culture in which people are encouraged to think freely and be curious in exploring new ways of working.

The best ideas for OSH innovation are generated by employees, and as an organisation there is always more you can do to tap into this reservoir of knowledge. Therefore, it is important to spend time considering the different ways you may be able to gather contributions from employees, such as using surveys and focus groups.

Change management

Organisations often have to change strategies, tactics and behaviour to realise improvement. There are various models to provide a structure for managing change in organisations. One of the most applicable to OSH practitioners is the eight-step process developed by John Kotter (Kotter, 2012). Kotter considers organisational change in the broadest sense, but the steps in his process apply to any OSH innovation, as detailed below:

1) Create a sense of urgency – there are various ways in which you can make OSH more important to decision-makers:

   • Provide objective data illustrating trends in workplace incidents
   • Support data with cost projections to illustrate the business impact of doing nothing
   • Empathise with the decision-makers and communicate the change required in a language that appeals to their emotions
   • Provide evidence and benchmarking information to demonstrate what the competition is doing in this area.

2) Build a guiding coalition – gain support from key people in your organisation at all levels. There will be influential people who are supportive of OSH change efforts; you need to bring them together.

3) Form a strategic vision and initiatives – make sure everyone understands what needs to change, why the change is important and how it will be achieved. All OSH change efforts need to be supported by a message that clarifies the reason for change. Take time to articulate a vision for OSH change and use a variety of mechanisms to communicate it to the organisation.

4) Enlist a volunteer army – form a large group of individuals to implement, monitor and maintain the change initiative. Although it is important to have key leaders to drive OSH innovation, you will also need enough people from different levels of the organisation who are ready and willing to support the change. You may be familiar with the theory of "social proof" which holds that if you can get enough people doing something others will follow (Cialdini, 2007). This phenomenon is particularly important in implementing OSH changes.

5) Enable action by removing barriers – address the people and processes that may get in the way of change. There will be people who are sceptical about OSH changes. Try to connect with them and address their concerns. If the barriers are structural or process-related, you may need the support of your change leaders to tackle them.

6) Generate short-term wins – motivate people through the realisation of short-term targets. When you introduce a new OSH vision its aspirational long-term goals may be exhausting for some people, so it is important to create sub-goals and to measure performance against them periodically. Passing milestones will help to create a sense of accomplishment and generate the impetus to improve further.

7) Sustain acceleration – try to maintain the momentum associated with the change effort. A variety of techniques can help to revitalise efforts:

- Reinforce the vision for change and communicate progress updates
- Introduce new members into the guiding coalition if necessary to re-energise the process
- Implement new projects and themes to keep the change effort fresh and in people's minds
- Reward and recognise people who are exemplars of the change.

8) Institute change – anchor the change in corporate culture by ensuring it becomes part of the new organisational behaviour. For change to become sustained reality, there should be the necessary systems, processes, people and environment. Training and communication will help in instituting the change but there is no quick-fix solution for meaningful organisational change; it requires persistence, resilience and unrelenting leadership commitment.

With the recent introduction of ISO 45001, now is a great time to be thinking about innovation. The Standard introduces various new concepts, such as understanding organisational context and opportunity management, which will require OSH practitioners to come up with new ideas to integrate these requirements within OSH policy and management arrangements. However, despite growing awareness that organisations need to embrace a culture of innovation and change for continual improvement, a dilemma faced by many OSH professionals is that organisations may be inclined to invest time and money on OSH improvements only when things go wrong.

Cynics may argue that if an organisation already has good OSH performance then there is no need to innovate and change on the principle that "if it ain't broke, don't fix it". Although there is truth to this idiom, there is also the danger that by standing still other organisations will pass you by. There are many examples in the business world of organisations which have suffered drastic consequences from stagnating and failing to innovate; Kodak is an obvious example (Mui, 2012). Fundamentally, there will always be opportunities for continual improvement in OSH management and practice that we should be seeking to explore.

The challenge is being able to convince decision-makers to be proactive in supporting change efforts when long periods of time may have passed without an OSH incident. Ultimately, it will be your skills as an OSH practitioner that will determine how effectively innovation and a willingness to change is embraced in your organisation. Remember the discussion in Step 2 on *General Management* that one of the key distinctions between leadership and management is the ability to generate change. Change, by definition, requires the creation of new systems and ways of working, which in turn always demands leadership, whereas management is typically more concerned with ensuring business functions operate efficiently and effectively.

 OSH practice involves setting direction and developing fresh approaches to OSH challenges, so make sure to set aside enough time for devising a strategy for increasing creativity and innovation in your approach.

Information and communication technology

Although technology is referred to in the OSH innovation typology, the risks and opportunities created for OSH practice from ICT merit further discussion. The workplace has seen a huge growth in the use of ICT, as well as rapid changes in the type of technology used, including computer networks, electronic data interchange and the internet. Many organisations have developed distinct ICT strategies and it is important that OSH is addressed within these strategies. If these types of strategies and associated plans are not in place, it may be worthwhile considering a distinct OSH strategic approach to ICT and the different ways in which ICT may be able to assist in accelerating improvements in OSH performance.

OSH software solutions

Many organisations are investing in OSH software solutions to assist with OSH management and there are an increasing number of proprietary OSH management solutions available on the market. OSH practitioners often get involved in the selection and implementation of OSH software systems, with the following considerations important in this process (Verdantix, 2014):

- User interface design – the software should be simple and easy to use as if the system is not intuitive then this will have a significant impact on user adoption
- Breadth of OSH workflows – to avoid having multiple separate systems, try to select software with multiple modules for specific OSH capabilities
- Degree of configurability – the ability to be able to manipulate workflows, make changes to forms and notifications helps to ensure the system is specific to your organisation
- Availability of mobile apps – increasingly, people want to use their smartphones and mobile devises when conducting inspections and audits
- Off-line capability – the ability to use the system without an internet connection can be useful when conducting site visits and inspections away from office environments, or in locations where data connectivity could be a problem
- Integration – quite often organisations may be using other enterprise-wide tools for risk management, making the ability for software systems to integrate a unique selling point
- Costs – the costs involved in procuring the license/subscription needs to be evaluated. It is also important to be aware of costs associated with implementation, including training that may be required to deploy the system effectively.

Sufficient time should be set aside during the procurement of any OSH software solution to gather references and feedback from people who have used the software. Sometimes OSH software vendors will provide customer references from specific industries, however, try to speak in person to representatives from similar organisations to gain their feedback on systems.

When procuring an OSH software solution gather references and feedback from people who have used the software. OSH vendors will provide customer references, but take time to speak in person to representatives from similar organisations.

Social networking

Another relevant example of the growth of ICT is the popularity of social networks. Although we need to be careful with social networks, particularly in terms of the information that is shared with others, there is no doubt about the power of social media in the context of OSH in communicating with an extended network of people who share similar interests. The following sections touch on some of the common social media tools and the benefits and limitations for OSH practitioners (Leathley, 2012).

LinkedIn

LinkedIn can be useful as a networking tool; however, like many social media tools, there is the risk of building up a substantial number of loose connections who you do not know, or fail to take the time to engage with and get to know. In this sense, it is worthwhile trying to find time to connect with people on LinkedIn by sending them a personal note and trying to keep in touch over time. LinkedIn also provides a useful way of getting in touch with people that you may have lost contact with during your profession. You never know how useful a connection from the past may be to your career progression! Increasingly, recruiters are using LinkedIn as a source for identifying candidates for OSH positions. Therefore, if you are looking for your next move it makes sense to include a comprehensive resume on LinkedIn so that potential recruiters can quickly gauge your skills and experience.

Twitter

Twitter is microblog that allows you to broadcast what you are thinking – micro as you are limited to a maximum of 280 characters. Its popularity stems from the fact that it is very easy to "tweet" from your mobile or personal device and that "followers" are not overwhelmed by information – in the age we live in of attention deficiency this can be very useful! There are numerous OSH organisations that regularly use Twitter to share useful headlines. A few examples of organisations with twitter feeds that might be worthwhile following include:

- The HSE: @H_S_E
- NEBOSH: @NEBOSHTweets

- IOSH: @IOSH_tweets
- IOSH Magazine: @IOSHmagazine
- IIRSM: @IIRSM
- The European Agency for Safety and Health at Work: @EU_OSHA
- OSHA/US Labor Department: @USDOL

Twitter can also be useful is developing your own personal brand and status with many high-profile OSH personalities regularly using Twitter to share stories and comments. To get to the widest audience, that is, people who are not yet following you, include the terms most relevant to the tweet preceded by a hashtag (#) so, #stress or #stressmanagement and #healthandsafety. So, if you have something interesting to say, it may be worth considering how you can use Twitter as a way of getting people to know who you are. However, be careful not to turn into the Donald Trump of OSH by tweeting too many messages of self-importance!

OSH forums

The Institution of Occupational Safety and Health (IOSH) make use of such networks, as evidenced by their OSH forums. These forums provide the opportunity for people to raise OSH questions. Questions can typically be addressed by OSH practitioners, who possess an understanding of the related subject matter.

If you are going to raise questions on a forum try to ensure that these are questions that can't be answered by a quick visit to the HSE website, or other relevant sources that can be accessed from the internet. Sometimes these types of questions can be damaging to your credibility. Also, try to ensure that questions are pointed and specific to the OSH matter in question. Unfortunately, it is not uncommon to read threads on OSH forums which have started with one issue but gradually resulted in a discussion that was completely tangential to the central subject matter.

The OSH entrepreneur

Have you ever thought about setting up your own company? I am sure that there must have been a time when you were sat at work, sick of being told what to do by your incompetent boss and yearning for the opportunity to try something different? There is no getting away from the fact that going out on your own is a significant risk, however, the potential benefits can be substantial.

If you do decide to set up your own business you will be part of a growing community of OSH entrepreneurs. Indeed, in the UK alone, a recent search of the Occupational Safety and Health Consultants Register (a UK web based directory of OSH consultants) showed that there are approaching 2,000 registered consultants offering OSH consultancy services (Occupational Safety & Health Consultants Register, n.d.).

Developing entrepreneurial opportunities

Successful entrepreneurs can transform great ideas into great businesses. However, the starting point is identifying good ideas that could become viable business opportunities. Why not spend some time thinking about the potential for monetising OSH opportunities? There are a few ideas below to get you started:

- Knowledge – specialist information, know-how or expertise, such as the provision of specialist OSH advisory services where there is a gap in the market
- Technology – the application of technology to solve OSH problems, meet needs or create new products or services
- Product – delivery of existing OSH products to meet market demand or incremental innovation to find new markets
- Service or experience – this could cover a wide range of activities to address actual or potential demand
- Social market – where people and communities have needs, generally for services, typically where state-based services are not functioning effectively
- Cultural, social or lifestyle – OSH opportunities could be linked to the fields of leisure, tourism, hospitality, culture and entertainment
- Physical resource – exploitation of land, water or naturally occurring resources
- Trading and commodity – based on buying and selling in relation to market conditions of supply and demand. There could be trading opportunities in OSH as we have seen in the environmental space.

So how did you do? Remember that entrepreneurship is not always about coming up with something entirely new. Often it can be tweaking or refinement of existing business models. Well, hopefully this got you thinking about entrepreneurial opportunities – the next step is to put these ideas into action!

Starting your own OSH business

The UK Government has created useful guidance on the key steps to starting your own business (GOV.UK, n.d.). These are discussed below in the context of the OSH profession:

1   Start with an idea – Embracing entrepreneurial spirit may help you in making the jump to self-employment, but make sure you can provide a product

or service that people will want to buy (see Step 7 on *Marketing and Brand Management* for further discussion on marketing strategy).

2   Get funding – Carefully consider how you intend to finance your business. A variety of options may be available, including self-funding through bank loans or equity funding. In recent years, crowd funding has also become popular, whereby a large group of people invest money in a business idea. In some countries, such as the UK, there are government schemes available to support new businesses.

3   Research your market – It is vitally important to determine whether there is a demand for your product or service. The market for OSH products and services can be very competitive, so spend time thinking about what is going to make you different, in terms of why consumers should select you over the competition.

4   Develop a plan – A compelling vision is important when starting your business. Your business vision should be supported by a comprehensive and robust business plan. The business plan should show the results of your market research and detail the specific arrangements for implementing your business strategy.

5   Find partners, suppliers and premises – The presence of a strong supply chain is critical to a successful business. Ultimately, you want to be looking for credible, consistent partners that you can rely on to provide an excellent service

6   Set up your business – Once the investment has been made and the planning completed it is down to you to execute your vision!

Summary

Innovation is synonymous with continual improvement and an important focus area for distinctive OSH practitioners. Ask yourself what you want to be remembered for and surely being different would be high up on the list. Try to consider innovation from a broad perspective and consider the different aspects of your organisation and its activities and explore opportunities for OSH innovation. Look to liaise closely with others in this process and learn about different ways of working and how OSH innovation can become part of your organisational culture.

If you are toying with the idea of starting your own OSH business, try to see how you can be distinctive in terms of the products or services that you offer. The more creative and unique your business offering is the greater chance you have of creating a successful and sustainable business and ultimately making a profit. There is no getting away from the risks involved in taking this step, however, the rewards might get you thinking why it took such a long time to make the decision!

Key takeaways

•   Innovation is the practice of making changes in something established, especially by introducing new methods, ideas or products with the creation of an

OSH typology an excellent way to identify opportunities for innovation – *have you developed on OSH typology to identify focus areas for OSH innovation?*

- An understanding of change management and its application can help to ensure that potential new ways of doing things quickly build momentum – *do you have a strategic approach to managing OSH change?*
- ICT is helping the world to become more interconnected and creating risks and opportunities for OSH practice – *does your approach to OSH management embrace ICT?*
- The growth of social media, such as LinkedIn and Twitter, provide excellent platforms for communicating OSH messages – *are you leveraging social networks sufficiently in the pursuit of your organisational and individual OSH goals?*
- Embracing entrepreneurial spirit can help in making the jump to self-employment, but make sure you have a compelling vision and plan before starting your business – *have you identified and realised opportunities to commercialise OSH?*

References

Budworth, T., & Ghanem, W. (2014). *Reflective learning: An essential tool for the self-development of health and safety practitioners*. London: Routledge, 1st edition.

Christensen, C. M. (2011). *The innovator's dilemma: The revolutionary book that will change the way you do business*. New York, NY: HarperBusiness, Reprint edition.

Cialdini, R. (2007). Social proof: Truths are us. In *Influence: The psychology of persuasion*. New York, NY: HarperCollins Publishers, 87–125.

Gersch, K. (2013). Google's best new innovation: Rules around '20% time'. *Forbes*. Retrieved from www.forbes.com/sites/johnkotter/2013/08/21/googles-best-new-innovation-rules-around-20-time/#4d39031c68b8.

GOV.UK. (n.d.). *Start you own business*. Retrieved from www.gov.uk/starting-up-a-business/set-up-your-business.

Kotter, J. (2012). *Leading change*. Cambridge, MA: Harvard Business Review Press.

Leathley, B. (2012). Only connect: Social media for safety professionals. *Health and Safety at Work Magazine*.

Mui, C. (2012). How Kodak failed. *Forbes*. Retrieved from www.forbes.com/sites/chunkamui/2012/01/18/how-kodak-failed/#50b01bc4bd6a.

Occupational Safety & Health Consultants Register (OSHCR). (n.d.). Retrieved from www.oshcr.org/.

Verdantix. (2014). *Green quadrant EH&S software*. Retrieved from http://enablon.com/reports-download/green-quadrant-ehs-software-global-2014-3juxca91cx.

10 Interpersonal skills

Well done – you've made it to Step 10! This is the final Step linked to last theme in the OSH Practitioner Transformation Model on the need to "be distinctive". The preceding Steps have provided you with underpinning knowledge of the theories, frameworks and tools taught in most business schools. This information will undoubtedly help you understand the broader concepts of business and how they relate to OSH; however, you will also require the necessary interpersonal skills to ensure that knowledge becomes power! In this Step, the focus will be on the concept of emotional intelligence and how you can develop your capability to persuade and influence decision making in your organisation.

On completion of Step 10, you will be able to:

- Explain what are interpersonal skills and their importance to your career progression
- Identify key interpersonal skills, including communication, teamwork, conflict resolution, empathy, networking and curiosity
- Recognise how to communicate impactful OSH messages, focusing on the key elements of compelling presentations
- Appreciate the value of creating a strong and meaningful professional network
- Understand power in organisations and the importance of defining your path to power as an OSH practitioner.

Does competency matter?

This may appear somewhat of a flippant question, as competency is undeniably important to your career and the impact that you have on OSH in your organisation. However, it may not be the key determinant to your career progression. We would all like to believe that our organisations are meritocracies, but this is not always the case—the most competent people, who work the hardest, are not necessarily those who are successful.

Unfortunately, the world is not always a fair place and if we want our organisations to achieve OSH goals then unfortunately we may have to play politics. Indeed, people who can coordinate actions in organisations tend to be those who understand how the system works and how to influence others. Increasingly, emotional intelligence is being cited as more important at work than competence (Goleman, 1996), so it may be worthwhile looking at developing your interpersonal skills. IOSH Blueprint – a self-assessment tool enabling OSH professionals to assess their competencies and produce professional profiles and development plans – also places substantial emphasis on the need for interpersonal skills (IOSH, 2016a).

What are interpersonal skills?

Interpersonal skills are the abilities that enable us to communicate and interact with others. In this Step, we will be exploring several skills including communication, teamwork, conflict resolution, empathy, networking and curiosity. Interestingly, one of the statements that received the highest score in our *OSH Practitioner Insight Survey* was on the importance of interpersonal skills, with OSH professionals strongly believing these skills are important for career progression.

 96% of respondents either "agreed", or "strongly agreed" that interpersonal skills are important for career progression.

Communication

Communication is a factor commonly cited as important in the development of a positive OSH culture, but how effective is your approach to OSH communication? Communication is a critical activity for OSH practitioners as people in an organisation may not always understand *what* you are doing and more importantly *why* it is important to the organisation. The belief shared by some people that OSH provides little or no added value to an organisation provides an even stronger motive to justify the business benefits of your activities on a regular basis.

A common practice for many OSH professionals is to focus OSH communication on loss events and other failures in OSH management arrangements. This alarmist approach is designed to jolt decision makers into action, but we need to ensure that our approach focuses on opportunities and not just threats. There will always be examples of bright spots within your organisation, where OSH initiatives have had a positive impact on business performance – these need to be showcased. For example, next time you hold in-company training, win an OSH award or receive positive feedback from OSH activities, make sure you capitalise on this opportunity and market the business benefits to a wide audience.

Presentations

Think about the last time that you attended a presentation when you were truly captivated. What was it about that presentation that created such a powerful impression? Every time that you stand up and deliver a presentation is a window of opportunity. It is important that you are adequately prepared for these opportunities, should it be a brief presentation to your team or a paper being delivered at a conference. In relation to interpersonal skills, the subject of public speaking and presenting has been extensively researched and the amount of published literature in this area is overwhelming. However, a few basic tips from the most successful TED (Technology, Education and Design) talks can help ensure that your presentations touch peoples' hearts, are unique and memorable (Gallo, 2014):

- Touch my heart

    - Ensure that you are inspired – make sure you are excited and passionate about the OSH subject when presenting
    - Speak to the heart – connect with people on a personal level and remember that the emotional (hearts) content in your presentation should outweigh the logical (minds) component
    - Understand the audience – tailor your material to the audience and practice intently so that you can internalise the presentation

- Teach me something new

    - Be different – introduce an emotionally charged event or create a "wow" moment, not another boring 5 steps to risk assessment presentation!
    - Offer a fresh approach to solve old problems – present existing OSH data in new ways
    - Don't be a brown cow! – be remembered for being unique and creative – a purple cow (Godin, 2009). If people are still talking about your presentation one month later then that is a success

- Present content in ways I'll never forget

    - Stick to the 18-minute rule – limit the duration of your presentation to keep people interested in the subject
    - Understand that thinking and listening is hard! – use humour to lower defences and incorporate analogies, metaphor, pictures and learning aids
    - Use the "Rule of 3's" – apply this approach to create memorable presentations (see Figure 10.1).

Emails

So much of our OSH communication takes place by email that it is necessary to spend some time discussing this communication tool. Sending an email is an

Rule of 3's

When designing presentations, the "Rule of 3's" can be useful in ensuring that your content is memorable. There are some great examples applicable to OSH that come in threes (e.g. stop, look and listen/clunk, click, every trip) which have stuck in minds of people. When developing your next presentation try to apply these principles in consideration of the following:

- What is the big idea from the presentation (i.e. the one thing that you want people to remember from the session)?
- Can the big idea be broken down into three related topics?
- Is there value in breaking down each topic into three sub-topics?
- Is there a summary that reinforces the big idea and the related topics?

Some of the best presentations have applied this approach, however, just a final word of caution. Although the "Rule of 3's" is useful, don't become a slave to this approach! There may be times when there are not three, but four, five or more topics and it is not appropriate to shoehorn this rule into all presentations.

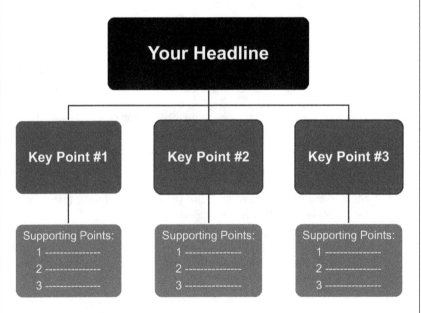

Figure 10.1 Framework for developing Occupational Safety and Health presentations

incredible opportunity to establish and reinforce your organisational and personal brand, so it is worthwhile considering what impression you are creating when you send emails. Be aware that although email is extensively used in most organisations there are various traps that we can fall into which limit its efficiency and effectiveness (Meyer et al., 2005). The following simple guidelines will help maintain a consistent and professional approach when communicating OSH messages:

- **Standardise email settings** – Configure standard font and colour for all emails within your outlook settings, in line with company brand guidelines. When emails are sent from people in the same business using different fonts, styles, sizes and colours this can provide an unprofessional image.
- **Use the "To" and "Cc" fields appropriately** – If you need someone to respond to an email, make sure they are included in the To: line; otherwise, if you just need people to be kept in the loop include them in the Cc (Carbon copy): line. Many people place all addresses in the To: line and never use the Cc: option. However, when an email is unintentionally sent to multiple people it can be unclear who the message and any associated actions are intended for. Ultimately, this can slow down the time taken to respond to the issue raised.
- **Maintain only one topic per email** – Try to keep each email exchange focused on one idea or one problem/solution. Communication can get confusing when too many topics get caught up in an email chain. You may have been brought into an email discussion which is covering a range of topics which can make it difficult to discern what is expected from you. If you need to discuss a different topic with the same person then send a new email with a new subject line.
- **Incorporate hyperlinks instead of attachments** – When you need someone in your organisation to access a file, where possible, insert a link to the file saved in a shared folder instead of attaching it. Sending shortcuts is a great way to reduce email traffic and ultimately reduce your mailbox size and the one of the receiver. Hyperlinks are quick and easy to establish by making a copy of the file path and then selecting the word(s) to contain the hyperlink. However, do make sure the receiver is able to access the file, as there may be occasions, for example sending emails to external parties, when this is not feasible.
- **Sign off emails appropriately** – Include a brief sign-off statement in every email. The sign-off (e.g. "Regards,") is an important part of completing an email. Your sign-off can be incorporated into your automated email signature, followed by your name, to help save time in writing this out every email.
- **Include your signature in every email** – Every time you send an email, whether internally or to external parties, make sure to include your email signature. Your email signature is one of the ways in which we can introduce

your company and personal brand. Subsequently, it is important that it is included in every email. Also, have you ever had to trawl through a trailing email to try and find someone's contact details? This way your contact details are readily available every time.

- **Flag up emails to be responded to later** – Assign flags to emails which you receive and intend to action later. If you receive an email which you know will require time to reply to, let the sender know you've received their message and that you'll respond more fully later. Also, incorporate an "Add Reminder" including a specific date and time to follow up the email.
- **Save emails appropriately** – Save emails in the relevant business folders and archive emails accordingly. An important part of good email management is saving associated emails in the correct folders. Emails that should be saved are those that will be of interest to other team members and those that have/might have significant business implications. Also, look to locally archive all your emails (both sent and received) so that you and your organisation have a record of all communications.
- **Use out of office replies** – When you are out of the office for a lengthy period ensure that an out of office reply is configured. It can be frustrating to send a message and then find out subsequently that the person is away from the office or on leave but has not set up an out of office reply. Your message should include the time of your return and a contact number or email address of a colleague who can respond if necessary in your absence. As this is often the last thing you might do before rushing off on holiday, check the message carefully for spelling mistakes before automatic replies are set for the account.

Teamwork

When devising and delivering OSH solutions you are often required to work as a team. The literature relating to team development and dynamics is extensive, with group development often presented as a series of stages known as forming, storming, norming and performing (Tuckman, 1965). A fifth stage of adjourning was subsequently added to the process, which is sometimes known as mourning (Tuckman & Jensen, 1977). This model has its shortcomings in that it is not clear how long each stage is likely to last, or what will prompt an evolution to the next stage of development, however, it is still regarded as a useful predictor of group development.

As so many OSH decisions are made by teams, it is important to be aware of the common traps that groups fall into:

- Social loafing – the tendency for some members of a group to "take a back seat"
- Group think – the group becomes overly focused on agreement and unanimous decision making at the expense of a realistic solution to the problem at hand

- Group polarization – the phenomenon whereby groups are inclined to make riskier decisions due to shared responsibility
- Group memory – the understanding that a group has better memory recall than the individual member with the best memory.

 When making decisions as a group, try to look at the subject matter objectively and ensure your decisions are not heavily influenced by the dynamics of the group.

Conflict resolution

Conflict is a regular and necessary feature of running any successful business; we cannot always expect people to agree on the best ways to improve performance. The same can be said of OSH, where disagreements will commonly be rife in terms of the most effective ways to change attitudes and behaviours in the workplace. Subsequently, the OSH practitioner needs to be skilled in the art of conflict resolution to ensure that conflict does not have a destructive impact.

How effective are you at managing conflict in your organisation? Not everyone is going to agree with your OSH approaches, due to differing goals, priorities and philosophies. Invariably, we are presented with difficult conversations and need to possess an armoury of tools to arrive at positive outcomes (Stone, 2010). A basic understanding of the following five conflict management strategies will help you deal more effectively with conflicts before they escalate beyond repair:

1 **Compromise** – giving up something and sharing in a compromised outcome with no winners or losers. Recognise the times when each party stands to lose something valuable and try to reach a compromise (e.g. agreeing on joint actions between different individuals/departments following an OSH incident, as opposed to apportioning blame).
2 **Competition** – dominating others to achieve your own goals. Resolution can be viewed as win/lose. There may be occasions where the gravity of the situation requires you to be more assertive (e.g. responding to emergency or crisis events).
3 **Collaboration** – satisfying the concerns of all parties and co-operating to achieve a mutually beneficial outcome (win/win). On contentious issues, look for opportunities for integrating ideas from multiple people (e.g. involving different functions in OSH policy/process development).
4 **Avoidance** – seeking to put off conflict indefinitely. By delaying or ignoring the conflict, the problem may resolve itself without a confrontation. This may

be more suited to low risk scenarios (e.g. avoiding escalation of trivial OSH issues).

5   **Accommodation** – willing to appease the other by placing their interests above your own. This is not necessarily a sign of weakness but may be a tactical manoeuvre – losing a battle but still intent on winning the war! (e.g. taking a soft line on certain OSH issues when you believe there are more important matters to fight for).

The most important factor in addressing conflict is for both parties to recognise that a problem exists and try to cooperate in solving it. Listening carefully to the views of others is also important in understanding the nature of the conflict. When dealing with more significant conflicts it is also important to adopt a structured approach to analysing the reasons for the conflict and identifying root causes and determining constructive and realistic solutions to dealing with the issue in hand.

 When faced with conflict ask the question "Why?" and keep asking until you uncover the root cause of the conflict.

Empathy

An important aspect of emotional intelligence is empathy, the ability to put yourself in someone else's shoes (Krznaric, 2015). As OSH practitioners, we need to identify the people and associated behaviours we need to change and speak in the language that will move these people. Invariably, different messages will need to be communicated to different audiences to reach out to their emotional and rational drivers. For example, when trying to convince top management to invest in OSH improvements, it would be advisable to focus on the reputational impact and cost savings that can be made from preventing occupational incidents and ill health. However, a similar approach may not yield success with front-line operatives. In this case, greater focus could be given to the impact on family members following a serious workplace incident.

Empathy involves trying to better understand an individual's point of view and it can be useful to lead with open-ended questions and statements during conversations to try and stimulate discussion:

*   Tell me about _____. By encouraging a person to talk by open questions you can start to ascertain some of the underlying reasons that could be central to a certain viewpoint
*   Why do you feel that way? The word "feel" is about emotions. It is important to connect with people at a personal level and understand emotional drivers
*   How can we do it better? This question suggests that the solution is a collaborative effort and may involve some level of compromise

- Help me understand the issue more clearly. The wording of this statement indicates that you perhaps are not as empathetic as is needed. It shows that you want to learn more about the other person's perspective.

These types of conversations enable individuals to share their perspectives as well as their suggestions and should lead to deeper levels of understanding. One of the problems in many organisations is busyness, where it appears that no-one has time to stop, talk and listen to their colleagues (Bruch & Goshal, 2002). Indeed, if a focus can be encouraged on developing meaningful conversations this can be an effective way of getting to know people better, caring about them and having a better chance of understanding matters from their perspective.

Next time you are in a meeting, listen carefully to the other attendees and write down the key factors driving their behaviour. When you pitch your ideas, try to do so in a way that will appeal to these factors.

Networking

Most people only think about networking when they are looking for a new job, but a strategic approach to networking can have a significant impact on your personal and professional development. Networking involves establishing and maintaining informal relationships with people whose acquaintance or friendship could bring advantages such as job or business opportunities. In its simplest form, networking is talking to people, becoming acquainted or friendly with them, and building relationships by getting to know more about them.

When meeting new people, we instinctively make decisions as to whether this person is beneficial to our personal and/or professional development. When you meet someone for the first time it is not always clear as to whether they could be a useful member of your professional network in the future, so try to be open to new connections – you never know when that contact may be of use!

One of the hardest parts of networking is also deciding what to do with people in your network. Perhaps you have 400 connections on LinkedIn, but what percentage of these individuals have you engaged with, and how often? If you are not remaining in contact with your network then what is the point in having the connection? There is no need to message everyone in your network every week or month, but you do need to establish some level of contact with your connections, clearly proportionate to the perceived level of the value individual contacts can offer. For example, you may want to send out an article, or guidance on an issue, or even reach out to individuals on their birthdays or during the holiday season.

Networking is difficult and doesn't come naturally to everyone, but as an OSH practitioner it can be an effective means of accessing useful information, learning about job opportunities and achieving your career goals. So, don't be reactive with your approach to networking by only using contacts when you get into trouble; try to embrace a more strategic and proactive approach to identifying and maintaining your professional network. Although you may not enjoy going out to social events, evening parties and other events held by your organisation, don't forget the power of informal organisations and the reality that many decisions are made away from its formal counterpart (Zack, 2010).

Curiosity

The final interpersonal skill to consider is curiosity. The importance of curiosity in leadership and management is a relatively recent development, however, it is increasingly recognised as just as important as intelligence (Chamorro-Premuzic, 2014). Essentially, curiosity relates to the level of interest that you have about your organisation and its inner workings. When things go wrong in an organisation do you want them fixed with no questions, or do you really want to understand what went wrong and why? Effective leaders tend to be curious about all aspects of their organisation and its ways of working.

Importantly, this 10 Step program emphasises the need to learn more about areas which are typically outside the remit of OSH practice. It encourages that you learn more about corporate and operational strategies and that you take time to understand the work carried out by different functions in your organisation. If you want others to be interested in OSH, then take time to understand what they do and the contributions they make to the organisation.

The next time you get an email about the financial performance in your organisation, about an IT project being deployed or a recent HR development, take time to read and absorb this information. Although it may not come naturally to everyone, try to be determined and disciplined in your approach to learn more about your organisation. In that way, when the time comes and you reach a senior level in your organisation you will be able to contribute to decisions not only on OSH, but wider matters.

Power in organisations?

Have you ever considered the importance of power within your role as an OSH practitioner? Well, if the answer is no, then perhaps it is about time that you did! The reason why is that powerful people can get things done, quite often by getting other people to do things for them, even things that others may not necessarily want to do. In the context of OSH, power is essential. If you are not able to influence others in the workplace to bring about change then you will find it difficult to generate continual improvement in OSH performance.

As an OSH practitioner, you may want to bring about change in your organisation. However, organisational change requires the ability to influence and persuade individuals at different levels. A commonly held misconception is that it is only those at the top of organisations that have the power to bring about change.

WORK 2022

In 2017, the Institution of Occupational Safety and Health (IOSH), the world's largest OSH membership organisation, launched WORK 2022 (IOSH, 2016b), a new five-year strategy aimed at shaping the future of the OSH profession. The strategy is founded on three key pillars:

1   Enhance – develop the safety and health profession
2   Collaborate – build strategic partnerships through collaboration
3   Influence – strengthen influence and impact globally

The WORK 2022 strategy emphasises that we must work collectively to address negative perceptions of OSH and provide OSH professionals, many of whom are working at strategic levels, with a broader range of skills and competencies.

To be effective, the OSH profession must continue to influence people regarding the positive role it plays in keeping people safe and well. As we progress to strategic levels in our organisations our interpersonal skills will take on increasing importance in achieving this goal and safeguarding the future of the OSH profession.

Indeed, most lessons on OSH culture and behavioural safety proclaim that top management commitment is the overriding factor in driving change management programmes. However, although top management commitment is important, power does not rest solely with top management. You may feel at times that you are unable to effect change as you have a lack of power. But power is out there for the taking. It is simply a case of understanding where power lies within your organisation and formulating a path to power (Pfeffer, 2010).

Summary

High levels of competency will not guarantee that you achieve your personal and professional goals. Often the most successful people know how to navigate the internal politics of organisations and employ interpersonal skills to influence and persuade others. This understanding does not infer a need to utilise Machiavellian or manipulative approaches; however, OSH practitioners do need to be better at using interpersonal skills to position OSH strategies within organisations to ensure a greater chance of success.

Power is a reality. Fundamentally, power is the ability to organise collective action – to get things done in organisations. Learning how to use power is of utmost importance to OSH practitioners, as although we are all working towards laudable

---

Power mapping

Have you ever thought about where the real power rests in your organisation? Well, one novel way suggested by Michael Segalla, Professor of Management at HEC Paris, is to survey managers and plot out your answers on a map to see the power patterns (Segalla, 2010).

To plot organisational power, try ranking individual managers on the following 3 dimensions:

- **Left Axis:** Sense of Responsibility – How much responsibility does the person have?
- **Right Axis:** Hierarchy – What level of seniority does the person have in the organisation chart?
- **Colour:** Objective Authority – Does the person have the authority to hire, fire and set salaries, budgets and strategies – or have a strong say in those decisions?

You might be surprised to find out that managers' authority and power is not always reflected by their positions in the organisational hierarchy.

This exercise will help you determine who are the key players in your organisation and ensure that you target your OSH initiatives effectively.

OSH visions everyone has different drivers and agendas and may not always be entirely committed to these goals. So, what are you waiting for? Embark on your path to power today, using interpersonal skills to achieve your professional goals and take your organisation's OSH performance to the next level!

Key takeaways

- Once competency is proven, your interpersonal skills will determine how far you progress in an organisation – *do you possess the necessary interpersonal skills to influence and persuade others?*
- Your interpersonal skills, such as communication, teamwork, conflict resolution, empathy, networking and curiosity, will be essential in achieving your personal and organisational OSH goals – *are you finding time to reflect on where developments in your interpersonal skills may be needed?*
- Remember to communicate with empathy and care and try to touch people's hearts, not only their minds – *are you delivering compelling OSH messages?*
- Building a strong professional network can be important in fulfilling your career goals – *have you adopted a strategic approach to building and maintaining your professional network?*
- Learn and be mindful of power in organisations and utilise the skill of power mapping – *have you identified the key players in your organisation and how to position OSH effectively?*

References

Bruch, H., & Goshal, S. (2002). Beware the busy manager. *Harvard Business Review*. Retrieved from https://hbr.org/2002/02/beware-the-busy-manager.

Chamorro-Premuzic, T. (2014). Curiosity is as important as intelligence. *Harvard Business Review*.

Gallo, C. (2014). *The 9 public-speaking secrets of the world's top minds*. New York, NY: St. Martin's Press.

Godin, S. (2009). *Purple cow: Transform your business by being remarkable*. Portfolio, New edition.

Goleman, D. (1996). *Emotional intelligence: Why it can matter more than IQ*. London, UK: Bloomsbury Publishing Plc.

Institution of Occupational Safety and Health (IOSH). (2016a). *IOSH blueprint*. Retrieved from www.iosh.co.uk/VP/Home/Toolkit/IOSH-Blueprint.aspx.

Institution of Occupational Safety and Health (IOSH). (2016b). *WORK 2022*. Retrieved from www.ioshwork2022.com/.

Krznaric, R. (2015). *Empathy: Why it matters, and how to get it*. New York, NY: Tarcher-Perigee.

Meyer, V., Sebranek, P., & Van Rys, J. (2005). *Writing effective e-mail: Practical strategies for strengthening electronic communication*. Burlington, WI: Upwrite Press.

Pfeffer, J. P. (2010). *Why some people have it and others don't*. HarperBusiness. 1st edition.

Segalla, M. (2010). Vision statement: Find the real power in your organisation. *Harvard Business Review*.

Stone, M. (2010). *Difficult conversations: How to discuss what matters most*. London, UK: Penguin Books. 10th anniversary updated edition.

Tuckman, B. W. (1965). Developmental sequence in small groups. *Psychological Bulletin*, 63(6), 384–399.

Tuckman, B. W., & Jensen, M. A. C. (1977). Stages of small group development revisited. *Group Organization Management*, 2(4), 419–427.

Zack, D. (2010). *Networking for people who hate networking: A field guide for introverts, the overwhelmed, and the underconnected*. San Francisco, CA: Berrett-Koehler Publishers.

Conclusions

So, what was the big idea behind this book? Well, from our experience to be taken seriously as an OSH practitioner and to have a more profound influence on OSH standards, culture and performance you must understand business. If you are an OSH professional looking to progress in your career to a senior level position then you can no longer just be a subject matter expert – you need to push the boundaries of your comfort zone and gain understanding in all aspects of your organisation.

As OSH practitioners who have been fortunate enough to have progressed into senior level OSH positions, largely through developing an increased knowledge of the world of business, it is incumbent upon us to share with you a word of caution. As you gradually transition into a senior level position you may find that you are involved in a wider range of business management activities (strategising, planning, budgeting, forecasting, reporting, etc.) and doing less of the things that may have attracted you to the OSH profession in the first place. An inevitable direction of travel, but still a direction that you should be aware of.

OSH as a profession and discipline is now firmly established. Furthermore, there is recognition that OSH must be embedded into the wider organisational strategies, processes and culture to generate improvements in OSH performance. For OSH to succeed it should become part of the fabric of organisations, no longer as a stand-alone function but something which is integral to the value proposition of organisations.

What does the future hold for OSH practitioners?

The introduction of ISO 45001 will undoubtedly bring about a step-change in the approach to OSH management. There will be greater focus on contextual factors, stakeholder management and OSH will become an integral aspect of strategy and leadership. This will require a transformation in OSH policy and practice to ensure that we are able to create and maintain meaningful and sustainable change.

The model and associated content presented in this book hopefully provide a useful framework for the self-evaluation and development of OSH practitioners in becoming more effective as OSH leaders and managers. We all want to succeed in what we do and whilst our passion for making the world a better place must be a driving force, we must also learn how to work smarter. Furthermore, it is in

Figure 11.1 Occupational Safety and Health practitioner transformation model

our hands to improve the perception and credibility of the OSH profession and to secure our own future.

This book by no means advocates that you should complete an MBA to bring about your transformation in OSH practice – however, you now have the skills to calculate the return on investment should you decide to go down this route! All the knowledge that you need to become more influential in the workplace is around you – it simply requires you to be curious in exploring all aspects of your organisation: its strategy, various functions and ways of working and ensuring that you, with your heightened awareness and interpersonal skills, position OSH in a way in which people will listen and act on your words.

So, we will leave you here to head back into your organisations to be more *strategic, cross-functional* and *distinctive* in your approach to creating a path to power for achieving your personal and professional goals.

Good luck! Rob Cooling and Waddah S Ghanem Al Hashmi

Annexure A

Occupational Safety and Health practitioner insight survey statements

1 I possess a good level of knowledge of the financial metrics used to measure performance in my organisation.
2 I fully understand the operations of my organisation.
3 I have a good understanding of the principles of marketing.
4 I have a good knowledge of operations management tools applicable to my organisation.
5 I can compile a robust business case for expenditure on safety and health.
6 I need to understand how the whole organisation works, in terms of all its functions, to succeed in my job.
7 The marketing function in my organisation plays an important role for safety and health.
8 To do my job well, I need to understand the marketing function in my organisation.
9 Understanding the finance and accounting function is critical for me as a safety and health practitioner.
10 Understanding investment appraisal is important in making a strong case for my organisation to invest in safety and health.
11 I have a good understanding of the distinction between leadership and management.
12 I am confident that I have instilled a compelling safety and health vision for my organisation.
13 I feel that as a safety and health practitioner, I am given a great deal of opportunity to learn about general management.
14 I am confident that I can think strategically on safety and health matters.
15 I believe that I possess the skills to create an effective governance framework for safety and health in my organisation.
16 I believe that my training and experience has prepared me sufficiently to become the next General Manager of my organisation.
17 I have a good knowledge of human resource management and its implication for safety and health management.
18 I possess a good understanding of how to design incentives to influence behaviours.

19 Corporate Social Responsibility is an important topic in my role as a safety and health practitioner.
20 Safety and health performance should be reported to shareholders in the same way that financial business metrics are reported.
21 Technical knowledge is important for me as a safety and health practitioner in influencing change in my organisation.
22 Innovation is very important to be successful as a safety and health practitioner.
23 Safety and health practitioners make good entrepreneurs.
24 Effective interpersonal skills are important for my career progression.
25 I have a good understanding of statistical methods and their application for the analysis of safety and health data.
26 I am creative in designing solutions for safety and health problems and challenges.
27 I understand how to communicate safety and health messages in a way to influence action.
28 I feel confident in my ability to empathise with others.
29 I have a good understanding of how to prepare reports on safety and health issues for management attention.
30 I believe that I am well respected in my organisation.

Annexure B

Cross-referencing the 10 Step MBA content to ISO 45001

Clause:	ISO 45001:	10 Day MBA Ref:
4	Context of the organization	
4.1	Understanding the organization and its context	General Management
4.2	Understanding the needs and expectations of workers and other interested parties	General Management
4.3	Determining the scope of the OH&S management system	General Management
4.4	OH&S management system	General Management
5	Leadership and worker participation	
5.1	Leadership and commitment	General Management Strategy and Leadership
5.2	OH&S policy	General Management
5.3	Organizational roles, responsibilities and authorities	General Management Organisational Behaviour and Human Resource Management
5.4	Consultation and participation of workers	General Management Organisational Behaviour and Human Resource Management
6	Planning	
6.1	Actions to address risks and opportunities	General Management
6.1.1	General	General Management
6.1.2	Hazard identification and assessment of risks and opportunities	General Management
6.1.3	Determination of applicable legal requirements and other requirements	General Management
6.1.4	Planning action	General Management
6.2	OH&S objectives and planning to achieve them	General Management
6.2.1	OH&S objectives	General Management
6.2.2	Planning to achieve OH&S objectives	General Management
7	Support	
7.1	Resources	General Management
7.2	Competence	General Management
7.3	Awareness	General Management
7.4	Communication	General Management

(Continued)

Clause:	ISO 45001:	10 Day MBA Ref:
7.5	Documented information	General Management
7.5.1	General	General Management
7.5.2	Creating and updating	General Management
7.5.3	Control of documented information	General Management
8	Operation	
8.1	Operational planning and control	General Management
		Operations Management
8.1.1	General	General Management
		Operations Management
8.1.2	Eliminating hazards and reducing OH&S risks	General Management
		Operations Management
8.2	Management of change	General Management
		Operations Management
		Innovation and Entrepreneurship
8.3	Outsourcing	General Management
		Operations Management
8.4	Procurement	General Management
		Operations Management
8.5	Contractors	General Management
		Operations Management
8.6	Emergency preparedness and response	General Management
		Operations Management
9	Performance evaluation	
9.1	Monitoring, measurement, analysis and performance evaluation	General Management
9.1.1	General	General Management
9.1.2	Evaluation of compliance	General Management
9.2	Internal audit	General Management
9.2.1	General	General Management
9.2.2	Internal audit programme	General Management
9.3	Management review	General Management
10	Improvement	
10.1	General	General Management
10.2	Incident, nonconformity and corrective action	General Management
		Innovation and Entrepreneurship
10.3	Continual improvement	General Management
		Innovation and Entrepreneurship

Annexure C

Cross-referencing the 10 Step MBA content to the NEBOSH Diploma syllabus (Nov 15) – Unit IA: Managing health and safety

Element	Element Title	Sub-element	10 Day MBA Ref
1	Principles of health and safety management	Reasons for health and safety management	Introduction
		Societal factors which influence an organisation's health and safety standards and priorities	Strategy and Leadership General Management Corporate Social Responsibility and Governance
		The uses of, and the reasons for introducing a health and safety management system	General Management
		Principles and content of effective health and safety management systems	General Management
2	Regulating health and safety	Comparative governmental and socio-legal models	Corporate Social Responsibility and Governance
		The purpose of enforcement and laws of contract	Corporate Social Responsibility and Governance
		The role and limitations of the International Labour Organisation in a global health and safety setting	Corporate Social Responsibility and Governance
		The role of non-governmental bodies and health and safety standards	Corporate Social Responsibility and Governance
3	Loss causation and incident investigation	Theories/models and use of loss causation techniques	Data Analytics and Decision Making
		The quantitative analysis of accident and ill-health data	Data Analytics and Decision Making

(*Continued*)

Annexure C (Continued)

Element	Element Title	Sub-element	10 Day MBA Ref
		Reporting and recording of loss events (injuries, ill-health and dangerous occurrences) and near misses	Data Analytics and Decision Making
		Loss and near miss investigations	Data Analytics and Decision Making
4	Measuring and reviewing health and safety performance	The purpose and use of health and safety performance measurement	General Management
		Health and safety monitoring	General Management
		Health and safety monitoring and measurement techniques	General Management
		Reviewing health and safety performance	General Management
5	Assessment and evaluation of risk	Sources of information used in identifying hazards and assessing risk	General Management
		Hazard identification techniques	General Management
		Assessment and evaluation of risk	General Management
		Systems failures and system reliability	General Management
		Failure tracing methodologies	General Management
6	Risk control	Common risk management strategies	General Management
		Factors to be taken into account when selecting risk controls	General Management
		Safe systems of work and permit-to-work systems	General Management
7	Organisational factors	Types of safety leadership and their likely impact on health and safety performance	Strategy and Leadership
		Benefits of effective health and safety leadership	Strategy and Leadership
		Internal and external influences	General Management Strategy and Leadership
		Types of organisations	General Management
		Third party control	General Management Operations Management

Element	Element Title	Sub-element	10 Day MBA Ref
		Consultation with workers	General Management Organisational Behaviour and Human Resource Management
		Health and safety culture and climate	Organisational Behaviour and Human Resource Management
		Factors affecting health and safety culture and climate	Organisational Behaviour and Human Resource Management
8	Human factors	Human psychology, sociology and behavior	Organisational Behaviour and Human Resource Management
		Perception of risk	Organisational Behaviour and Human Resource Management
		Human failure classification	Organisational Behaviour and Human Resource Management
		Improving individual human reliability in the workplace	Organisational Behaviour and Human Resource Management
		Organisational factors	Organisational Behaviour and Human Resource Management
		Job factors	Organisational Behaviour and Human Resource Management
		Behavioural change programmes	Organisational Behaviour and Human Resource Management
9	The role of the health and safety practitioner	The role of the health and safety practitioner	General Management Strategy and Leadership Interpersonal Skills
		The importance of effective communication and negotiations skills when promoting health and safety	Interpersonal Skills Marketing and Brand Management
		The health and safety practitioner's use of financial justification to aid decision making	Economics and Financial Management

Index

Page numbers in *italic* indicate a figure and page numbers in **bold** indicate a table on the corresponding page.